Felix Vüllers

Bioinspired Superhydrophobic Nano- and Microstructured Surfaces for Drag Reduction and Optoelectronics

Schriften des Instituts für Mikrostrukturtechnik
am Karlsruher Institut für Technologie (KIT)
Band 38

Hrsg. Institut für Mikrostrukturtechnik

Eine Übersicht aller bisher in dieser Schriftenreihe
erschienenen Bände finden Sie am Ende des Buchs.

Bioinspired Superhydrophobic Nano- and Microstructured Surfaces for Drag Reduction and Optoelectronics

by
Felix Vüllers

KIT Scientific Publishing

Dissertation, Karlsruher Institut für Technologie
KIT-Fakultät für Maschinenbau

Tag der mündlichen Prüfung: 17. Mai 2018
Referenten: Priv.-Doz. Dr. Hendrik Hölscher
Prof. Dr.-Ing. Bettina Frohnapfel
Prof. Dr. Chiara Neto

Impressum

KIT Scientific Publishing

Karlsruher Institut für Technologie (KIT)
KIT Scientific Publishing
Straße am Forum 2
D-76131 Karlsruhe

www.ksp.kit.edu

Print on Demand 2018 – Gedruckt auf FSC-zertifiziertem Papier

ISSN 1869-5183
ISBN 978-3-7315-0816-8
DOI 10.5445/KSP/1000084178

Bioinspired Superhydrophobic Nano- and Microstructured Surfaces for Drag Reduction and Optoelectronics

Zur Erlangung des akademischen Grades eines

DOKTORS DER INGENIEURWISSENSCHAFTEN

von der Fakultät für Maschinenbau des

Karlsruher Instituts für Technologie (KIT)

genehmigte

DISSERTATION

von

Felix Vüllers

Tag der mündlichen Prüfung:	17.05.2018
Referent:	Priv.-Doz. Dr. Hendrik Hölscher
Korreferentin:	Prof. Dr.-Ing. Bettina Frohnapfel
Korreferentin:	Prof. Dr. Chiara Neto

Abstract

Mimicking natural organisms and their unique properties, like superhydrophobicity, enables researchers to equip artificial surfaces with functionalities such as underwater air retention or self-cleaning properties. These functionalities can help to address challenges in a multitude of applications including reduction of fluidic drag in microfluidics and shipping or minimizing soiling of optoelectronic devices.
Inspired by the superhydrophobic leaves of *Salvinia cucullata* and *Pistia stratiotes*, which are covered in a dense layer of transparent hairs, this thesis introduces a highly flexible, superhydrophobic transparent thin nanofur film. Polymeric thin nanofur is fabricated using a combination of highly scalable hot embossing and hot pulling techniques. The resulting superhydrophobic film has extreme wetting properties with droplets rolling off the surface at minimal tilt angles and is self-cleaning against a broad range of contaminant species and sizes. The underwater retained air layer of thin nanofur, also called plastron, shows a good stability against hydrostatic pressure. In order to precisely control the retained air pressure and support the underwater retained air layer, thin nanofur is perforated and attached to a pressure support system, thus, significantly increasing the retained air layer stability against hydrostatic pressure and pressure fluctuations. By introducing a contact line motion measurement technique to analyze the dynamic plastron stability and contamination resistance of thin nanofur, several dynamic plastron stability enhancing topographical features are identified.
In addition to retaining an air film when submerged underwater, thin nanofur also exhibits self-cleaning and anti-reflective properties, making it applicable as a transparent coating and as a translucent self-standing film ideal for light managing

applications. Hence, superhydrophobic self-cleaning thin nanofur was successfully applied as an efficiency enhancing coating on organic light emitting diodes and solar cells. Moreover, based on the optically enhancing nanofur topography, a highly flexible superhydrophobic microcavity array is fabricated from mechanically, chemically and UV-stable fluorinated ethylene propylene typically used in the photovoltaic industry. Besides its anti-reflective and self-cleaning properties, the superhydrophobic microcavity array exhibits a high stability against impacting rain droplets, making it an ideal candidate for outdoor applications. These properties can be exploited by attaching the microcavity array to photovoltaic modules, and, thus, significantly increasing the modules electrical gain, while simultaneously equipping them with self-cleaning properties.
The scalable, highly flexible, superhydrophobic, self-cleaning, anti-reflective, air-retaining films introduced in this work present promising solutions for current challenges in optoelectronic and drag-reducing applications.

Zusammenfassung

Durch Nachahmen natürlicher Organismen und ihrer einzigarten Eigenschaften, wie z.B. Superhydrophobizität, gelingt es Wissenschaftlern künstliche Oberflächen mit Funktionalitäten wie Selbstreinigung oder dem Aufrechthalten einer Luftschicht unter Wasser auszustatten. Mit Hilfe dieser Funktionalitäten können eine Vielzahl von Herausforderungen in den verschiedensten Anwendungsfeldern, wie z.B. der Reibungsreduktion in der Mikrofluidik und Schifffahrt, sowie die Verminderung von Verschmutzung optoelektronischer Geräte, bewältigt werden. Inspiriert von den superhydrophoben Blättern der Wasserpflanzen *Salvinia cucullata* und *Pistia stratiotes*, welche von einer dichten Schicht aus transparenten Haaren bedeckt sind, wird in dieser Dissertation ein hochgradig flexibler, superhydrophober und transparenter dünner Nanopelz vorgestellt. Der aus Polymer gefertigte dünne Nanpelz wird mittels eines zweistufigen skalierbaren Verfahrens, bestehend aus Heißprägen und Heißziehen, hergestellt. Der so hergestellte superhydrophobe Film weißt extreme Benetzungseigenschaften auf, sodass darauf abgesetzte Wassertropfen bereits bei minimaler Neigung des Films abrollen. Des Weiteren weißt der Film eine selbstreinigende Wirkung gegenüber einem breiten Spektrum an Verschmutzungen auf. Die vom Nanopelz unter Wasser aufrechterhaltene Luftschicht, auch bekannt als Plastron, zeigt eine gute Stabilität gegenüber hydrostatischem Druck. Um eine genaue Kontrolle über den Luftdruck innerhalb der Luftschicht zu erhalten und sie gegenüber hydrostatischem Druck zu stabilisieren, wird der dünne Nanopelz perforiert und an ein Drucksystem angeschlossen. Hierdurch kann die Plastronstabilität gegenüber erhöhtem Wasserdruck und Wasserdruckfluktuationen signifikant erhöht werden. Durch die Verwendung einer neu eingeführten Methodik zur

Vermessung der Kontaktlinie, wird die dynamischen Plastronstabilität sowie die Widerstandsfähigkeit des dünnen Nanopelz gegen Verschmutzung analysiert. Dies ermöglicht die Identifikation mehrerer topographischer Merkmale, die zu einer erhöhten dynamischen Plastronstabilität beitragen.

Neben der Fähigkeit unter Wasser eine Luftschicht aufrechtzuerhalten, ist der dünne Nanopelz selbstreinigend und unterdrückt die Reflektion von einfallendem Licht, was seine Verwendung als transparente Beschichtung sowie als transluzenten frei stehenden Film ermöglicht und ihn zu einem idealen Material für Anwendungen im Lichtmanagement macht. Durch Ausnutzen dieser optischen Eigenschaften konnte der superhydrophobe, selbstreinigende Nanopelz erfolgreich als effizienzsteigernde Beschichtung auf Solarzellen und organischen Leuchtdioden angebracht werden. Außerdem konnte, basierend auf der optisch vorteilhaften Nanopelztopographie, ein hochgradig flexibles, superhydrophobes Mikrokavitätenarray aus mechanisch, chemisch und UV stabilem Fluorethylenpropylen hergestellt werden. Auf Grund seiner hohen Stabiltät gegenüber auftreffenden Regentropfen, sowie seiner antireflektiven und wasserabweisenden Eigenschaften, ist das hergestellte Mikrokavitätenarray ideal für den Einsatz im Freien geeignet. Durch Anbringen auf Photovoltaikmodulen konnte die Effizienz der Module signifikant gesteigert werden, während sie gleichzeitig mit selbstreinigenden Eigenschaften ausgestattet wurden.

Der in dieser Arbeit vorgestellte, hochgradig flexible, stark wasserabweisende, selbstreinigende, antireflektive und unter Wasser Luft haltende dünne Nanopelz stellt eine vielversprechende Lösung für eine Vielzahl von aktuellen Herausforderungen in Bereichen der Reibungsreduktion und der Optimierung optoelektronischer Geräte dar.

Contents

1 Introduction

The abundance of fascinating micro- and nanostructured surfaces in nature has always been a well of inspiration for researchers and engineers. Survival of the fittest and the need to find nourishment resulted in natural organisms and their surfaces being perfectly adapted to fulfill a variety of different purposes essential to survive in the species chosen environment. Among these functional surfaces are *Lotus* leaves [1], leaves of the water fern species *Salvinia* [2] and the water lettuce *Pistia Stratiotes* [3], which are equipped with special wetting properties. Their superhydrophobicity results in self-cleaning behavior necessary for unobstructed photosynthesis and enables them to retain an air layer when submerged underwater [4, 5]. The self-cleaning and air retaining properties observed on these natural surfaces are especially beneficial for technical applications including anti-fouling [6], self-cleaning windshields [7], solar panels [8], energy conservation [9] and drag reduction [10–13].
Developing efficient drag-reducing surfaces is of great interest to applications in medicine, shipping and fluid transport [14–16]. The integration of drag-reducing surfaces can reduce pressure loss and operational costs of microfluidic devices, which are essential for medical diagnostics, pharmaceutical, clinical and biological research [17–20]. Additionally, stable and scalable drag-reducing surfaces can help to reduce the enormous fuel consumption of international shipping, which has increased from 64.5 million metric tons (Mt) in 1950 up to more than 330 Mt in 2007 [21, 22]. Several techniques including microbubble injection, microstructuring surfaces with shark skin inspired riblets or adding polymer additives to the fluid have been employed to reduce frictional drag [14, 23–26]. However, limited efficiency, high production costs or the constant energy input required

to generate bubbles, limit these methods [14, 27]. In contrast, the passively retained air layer on natural superhydrophobic surfaces found on the skin of such insects as the water bug *Notonecta glauca* and the water strider *Gerris remigis* has a lubricating effect and enables them to efficiently move in and on water [10, 28]. Artificial air retaining superhydrophobic surfaces, such as polymers with functionalized zinc-oxide nanoparticles [29], etched and anodized aluminum surfaces coated with fluorinated chemicals [30–32] and fluorinated hydrophobized silicon microstructures [33], have adapted this lubricating effect and were shown to reduce frictional drag by up to 70% [26]. However, many of the fabrication techniques are cost-intensive and time-consuming, thus, making them difficult to scale up. Additionally, organic materials and micro- and nanosized particulate used for chemical surface treatment can be toxic to the environment, severely limiting the applicability of the functionalized surfaces. Another limitation is the instability of the retained air layer against external stimuli such as pressure and flow, which can replace the lubricating air layer with water and result in a loss of drag-reducing properties.

Superhydrophobic surfaces are also of great interest because of their self-cleaning properties. In particular, for optoelectronic applications, such as photovoltaic (PV) modules they present a way to reduce contamination. All around the year, natural aerosols such as sand, clay, volcanic ashes or pollen as well as anthropogenic pollution from power plants, vehicles and industry accumulate on PV modules [34–36]. These contaminants increase unwanted reflection and absorption of solar radiation at the protective encapsulation of PV modules, and, thus, reduce transmission of light into the active medium, leading to an up to 50% reduced energy output [34–40]. Geographical regions most prone to soiling and obstruction are arid regions, where rainfall is low and the possible electrical gain of PV modules especially high [34–40]. A variety of active cleaning mechanisms, such as manual cleaning, overnight storage or electrodynamic screens, can mitigate these effects, but also greatly increase the operational costs of the PV devices, thus, reducing their economic value [34–38, 41]. Therefore, the passive self-cleaning effect of superhydrophobic coatings is the most promising approach to solve the contamination problem, as it drastically reduces maintenance costs compared to active mechanisms, and satisfies most commercial requirements [34]. A variety of superhydrophobic transparent surfaces have been developed, and the most common of them are inspired by the anti-reflective coating of the moth eye and the transparent wings of the glasswing butterfly and the cicada [7, 8, 42, 43]. These coatings typically consist of subwavelength structures such as random nanopillars (glasswing) or nanonipples (moth eye) [8, 43], and are transferred onto optoelectronic devices through techniques such as etching, laser writing, two-photon lithography and soft lithography [42, 44, 45]. The fabrication of such transparent superhydrophobic surfaces often requires expensive and time-consuming fabrication techniques or subsequent chemical treatment, which makes upscaling difficult and limits their applicability [46–48]. Thus, the development of a safe, stable, biocompatible, easily scalable, self-cleaning, transparent drag-reducing material is of utmost importance.

The biomimetic group at the Institute of Microstructure Technology (IMT) has previously developed a bioinspired, superhydrophobic nanofur film [49–51]. This nanofur was shown to retain an air layer when submerged underwater, reduce fluidic drag and can be used as a liquid trap or an oil-water separating membrane [50, 51]. Based on these results, this work aims to improve nanofur and its properties, in order to make it applicable in drag-reducing and optoelectronic applications. Therefore, the objectives of this thesis are:

1. Development of a scalable and cost efficient fabrication technique for transparent superhydrophobic thin nanofur.
2. Equipment of thin nanofur with a highly stable underwater retained air layer.
3. Application of self-cleaning thin nanofur and nanofur-based films as efficiency enhancing coatings on optoelectronic devices.

To achieve these objectives, the following Chapter 2 presents the theoretical background regarding wettability, as well as fundamentals on the utilized fabrication techniques and analytical methods. Chapter 3 introduces the developed advanced fabrication and modification techniques used to fabricate thin nanofur and thin nanofur-based microcavity arrays and presents their respective wetting behavior. Next a characterization and subsequent improvement of the static and dynamic stability of the underwater retained air layer of thin nanofur against external stimuli is undertaken in Chapter 4. In Chapter 5 the optical properties of thin nanofur and thin nanofur-based microcavity array are analyzed and their efficiency enhancing effect on optoelectronic devices is demonstrated. Finally, Chapter 6 gives a summary and an outlook on possible future developments.

2 Theoretical and Methodological Background

2.1 Wetting Theory

When a liquid droplet comes in contact with a solid surface the two surfaces interact. This surface interaction most often results in the droplet spreading on the solid surface and forming a spherical cap, which is slightly deformed due to gravity. The extend, to which the droplet spreads on the surface depends on the interaction between liquid and surface. Wetting theory describes these interactions and their results (Fig. 2.1). Molecules in a liquid droplet are bound to each other by intermolecular forces, which, for molecules in the center of the droplet, are balanced in respect to all possible spatial directions. In contrast, for molecules residing in the outer ridges of the droplet, a higher number of molecules is positioned towards the center of the droplet than towards the edge, resulting in a net force pulling the outer molecules towards the center of the droplet. Thus, in order to move a molecule from the center of the droplet to the less bound outer region, an energy input is required to overcome the corresponding gap in potential energy (center = lower potential energy; outer shell = higher potential energy) [52]. Moreover, because the droplet is in contact with a surrounding vapor, solid or other liquid, molecules inside the droplet additionally experience forces from the environment, thus, changing the work necessary for displacement. The interfacial surface tension (γ) is defined as the work necessary for displacement per unit surface and describes the tendency of a liquid to minimize its surface area.

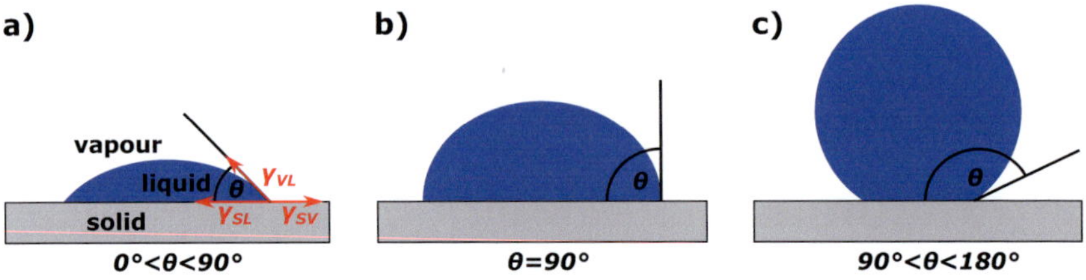

Figure 2.1: Schematic of different wetting behaviors. The contact angle at the solid-liquid-vapor triple line is defined by Young's equation (Eq. (2.1)) through the respective interfacial tensions and characterizes the wetting behavior of a surface. Surfaces with contact angle $0^\circ < \theta < 90^\circ$ are hydrophilic (a), while surfaces with contact angle $90^\circ \leq \theta \leq 180^\circ$ are hydrophobic (b ,c).

In 1805, Young [53] derived a relation between the contact angle (θ) of a droplet on a smooth surface and the surface tensions γ_{IJ}, of the liquid-vapor (LV), liquid-solid (LS) and solid-vapor (SV) interfaces, using a three-phase model and assuming a droplet small enough for gravity to be negligible.

$$\cos\theta = \frac{\gamma_{SV} - \gamma_{SL}}{\gamma_{LV}} \tag{2.1}$$

Equation (2.1) can be geometrically derived from Fig. 2.1 a) by regarding θ as a local property independent of drop volume and fixed at the triple line of the three phases. Because the contact angle (CA) is fixed and each surface tension tries to minimize the corresponding surface area, a balance of forces between the surface tensions immediately results in Eq. (2.1). An alternative derivation is based on minimizing the total surface energy (E) of a droplet $E = \Sigma_{LV}\gamma_{LV} + \Sigma_{SL}\gamma_{SL} + (\Sigma_S + \Sigma_{SL})\gamma_{SV}$, where Σ_{IJ} are the respective surface areas, and assuming a constant drop volume ($dV = 0$).

Based on the contact angle, wetting properties of solid surfaces are separated into different cases as illustrated in Fig. 2.1. Surfaces with a contact angle smaller than 90° are classified as hydrophilic and dispensed droplets wet the surface. **Super-hydrophilicity** describes an extreme case of hydrophilicity, in which the liquid exhibits a water contact angle $\theta_W < 10^\circ$ and completely spreads over the surface

[54]. Glass and most non-fluorinated polymers are hydrophilic with contact angles below 90°. In contrast, surfaces with $\theta_W > 90°$ are classified as hydrophobic, which in the extreme case of droplets exhibiting $\theta_W > 150°$ combined with low droplet adhesion is classified as **superhydrophobic**. To achieve contact angles above 90° surfaces are typically fluorinated or otherwise chemically modified, reducing their surface energy and increasing hydrophobicity. However, smooth surfaces cannot exceed a contact angle of 120°, even when chemically modified [55, 56]. To surpass this limitation and achieve higher contact angles, surfaces need to exhibit topographical features such as roughness or other regular or random surface structures [54, 57–59].
However, Eq. (2.1) only applies to chemically homogeneous, unstructured surfaces. Therefore, structured surfaces need to be investigated individually and two separate wetting states can be distinguished: Wenzel and Cassie-Baxter (Fig. 2.2) [4, 60–62].

2.1.1 Wetting States

Wenzel State

The introduction of surface roughness or surface features does generally alter the wetting behavior in one of two possible ways schematically shown in Fig. 2.2. The first case is known as "Wenzel wetting". A droplet resting on a structured surface penetrates into the topography and completely wets the surface. By introducing a roughness parameter r, which is defined as the ratio of actual surface area to the projected area of the surface, Wenzel was able to correlate the apparent CA (θ_{Wenzel}) of the droplet on the structured surface to the Young CA (θ_O) the droplet would have on an unstructured solid of the same material [60].

$$\cos\theta_{\text{Wenzel}} = r\cos\theta_O\,;\; r \geq 1 \tag{2.2}$$

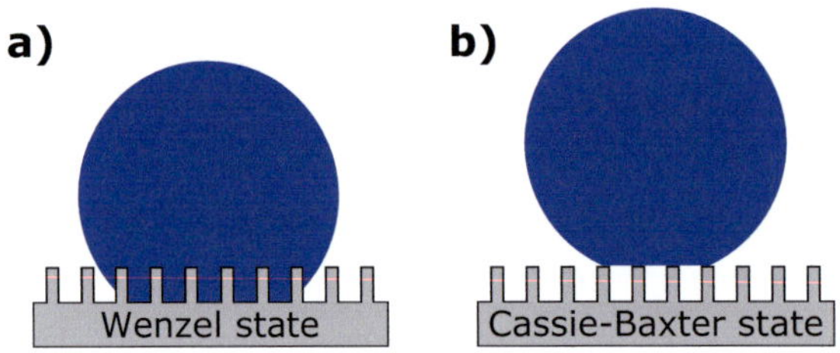

Figure 2.2: Schematic illustration of different wetting states. a) A droplet penetrating into the surface topography and completely wetting the structured substrate (Wenzel state). b) A droplet resting on top of the surface topography and trapping air pockets underneath (Cassie-Baxter state).

Because $r \geq 1$ surface roughness and topographical features always enhance the prevalent wettability of the bulk material. Therefore, a structured hydrophilic surface will exhibit lower contact angles than an unstructured substrate of the same hydrophilic material, while a hydrophobic material ($\theta_W \geq 90°$) will become more hydrophobic by introducing roughness [54, 57, 62, 63].

Cassie-Baxter State

The second possible interaction between a droplet and a structured surface is known as "Cassie-Baxter wetting". This state describes a droplet sitting on the apex of the surface topography without penetrating it, and, thus, partially resting on air pockets trapped in the structure underneath the liquid (Fig. 2.2 b) [55, 58, 64]. The air pockets reduce the liquid-solid contact area and significantly increase the contact angle of the surface. In 1944 Cassie and Baxter [61] developed a model to calculate the apparent CA (θ_{CB}) of these so-called "fakir droplets":

$$\cos\theta_{CB} = \phi_S r_S \cos(\theta_{OS}) + \phi_{\text{air}} r_{\text{air}} \cos(\theta_{O\text{air}}) \tag{2.3}$$

where ϕ_S and ϕ_{air} are area fractions of wetted solid and air respectively and r_S and r_{air} are the respective roughness factors of the wetted solid and air. Hence, the apparent CA in a Cassie-Baxter wetting state is an average of the CA of the two supporting materials weighted with the respective liquid-material surface area fractions. In the specific case of a fakir droplet partially resting on air, Eq. (2.3) can be simplified using $\theta_{O\text{air}} = 180^{\circ}$, $r_{\text{air}} = 1$ and $\phi_{\text{air}} = 1-\phi_S$:

$$\cos\theta_{CB} = -1 + \phi_S(r_S\cos(\theta_{OS}) + 1) \tag{2.4}$$

Thus, the CA in a Cassie-Baxter wetting state depends on the solid-liquid contact area and decreasing ϕ_S increases the CA. Furthermore, surfaces in a Cassie-Baxter wetting state often exhibit higher CA and at the same time significantly lower droplet adhesion than surfaces in a Wenzel state.

2.1.2 Contact Angle Hysteresis

High CAs are one prerequisite of superhydrophobic surfaces and can be achieved in both before mentioned wetting states. However, in order to be in a superhydrophobic Cassie-Baxter state, droplets additionally need to readily roll off the surface, which is not guaranteed by a high static CA. For example, droplets dispensed on rose petals and gecko feet display contact angles above 150° but firmly stick to the surface even when turned upside down [65]. Hence, to characterize droplet mobility and superhydrophobicity an additional physical quantity needs to be studied. Commonly, the roll-off angle or sliding angle (θ_{WS}) and contact angle hysteresis (CAH) are used to measure droplet adhesion and further characterize the wetting behavior of a surface. The roll-off or sliding angle is defined as the angle a surface has to be tilted for a liquid droplet to roll off the surface. Surface defects, chemical or mechanical inhomogeneities and roughness of the surface can result in variations in CA and allow a droplet to be pinned to the surface [54]. These differences in CA can be observed on droplets sitting on or sliding off a tilted plane. As gravity shifts the droplet down the surface, pinning results in a significantly higher observed CA at the front of the drop (advancing contact angle (θ_{adv})) than at the back of the drop (receding contact angle (θ_{rec})). This difference in CA is defined as contact angle hysteresis and can generate capillary forces large enough to compensate gravity and pin droplets to the substrate [66]. The large liquid-solid contact area in the Wenzel state usually results in droplets firmly sticking to the surface, and, thus, exhibiting high CAH and θ_{WS}. In contrast, a droplet in the Cassie-Baxter state is characterized by minimal CAH and roll off angles below 10°. This decreased droplet adhesion is a result of the reduced liquid-solid contact area combined with a "smoothing" effect of the air film covering defects and reducing roughness, thus, preventing Wenzel penetration [54, 63].

2.1.3 Critical Pressure

When a superhydrophobic surface is submerged underwater it can retain an air layer, as long as it is in the Cassie-Baxter wetting state. This is a result of the air pockets trapped underneath the liquid layer in between the surface topography (Fig. 2.2 b). Otten and Herminghaus [67] have identified roughness and elasticity of topographical features as key properties for surfaces to be able to retain an air layer and Solga *et al.* [68] expanded this list to include micro- to millimeter long hairs, hierarchical structures and cavities. Many such air retaining surfaces can be found in nature, including the water strider *Gerris remigis* [28, 69], the water bug *Notonecta glauca* [10] and water plants like *Salvinia* [11, 27, 70, 71] and *Pistia stratiotes* [72]. On biological surfaces the retained air film is used for a variety of applications such as drag reduction, respiration and insulation [2]. However, the retained air layer, which is essential for applications, can easily be damaged or completely broken by external stimuli such as immersion time or elevated hydrostatic pressure. The hydrostatic pressure necessary to induce a wetting transition from a Cassie-Baxter to a Wenzel wetting state is called critical pressure (p_{cr}) and can be calculated based on a balance of forces between hydrostatic pressure (p_{water}), ambient air pressure (p_{amb}), capillary pressure (p_{cap}) and retained air pressure (p_{air}) [73].

$$p_{air} + p_{cap} - p_{water} - p_{amb} = 0 \tag{2.5}$$

Several numerical and theoretical studies have shown wetting transitions to occur through two possible routes: sagging and sliding (Fig. 2.3 a) [73–76].

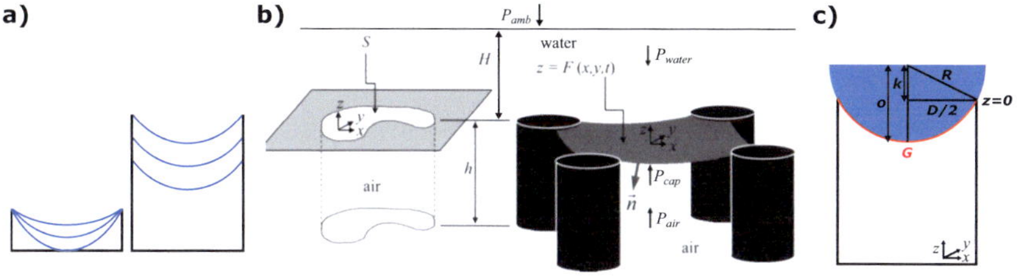

Figure 2.3: Schematic illustration of balance of forces on air-water interface. a) Schematic illustration of two possible wetting transitions: sagging (left) and sliding (right) b) Pressures acting on air-water interface inside pore (left) and between pillars (right) on superhydrophobic surface. (Reproduced with permission from Hemeda *et al.* [73]. Copyright 2014 American Chemical Society.) c) Schematic illustration of air-water interface (G) inside circular pore of diameter D. Water forms spherical cap with radius R inside pore.

First the air-water interface sags between the supporting structures until it reaches a critical curvature. Should the air-water interface contact the solid substrate during sagging, the surface undergoes an immediate wetting transition into the Wenzel state and the air layer is lost. If the air-water interface reaches its critical curvature without wetting the surface, it starts sliding into the topography while keeping the critical curvature, and upon contacting the solid substrate immediately wets the surface [73, 77]. In contrast to most theoretical studies investigating wetting transitions, which neglect the entrapped air pressure as stabilization mechanism for the air-water interface, Hemeda *et al.* [73] derived an equation for the critical pressure considering all involved pressures. Therefore, the presented theoretical considerations and calculations leading to an equation for calculating critical pressure follow the outline of Hemeda *et al.* [73]. First the superhydrophobic surface is assumed to consist of pores or pillars and the interface profile is $z = F(x,y,t)$ with $\hat{n}_F = (F_x, F_y, -1)/(1+F_x^2+F_y^2)^{1/2}$ being the unit normal as shown in Fig. 2.3 b). Capillary pressure is defined as the product of surface tension (γ) and divergence of the unit normal.

$$\begin{aligned} p_{cap} &= \gamma \nabla \hat{n}_F \\ &= \gamma \left[\left(1+F_y^2\right) F_{xx} + \left(1+F_x^2\right) F_{yy} - 2F_xF_yF_{xy} \right] \left(1+F_x^2+F_y^2\right)^{-3/2} \end{aligned} \tag{2.6}$$

Next the pressure in the retained air layer during compression is calculated. For this an isothermal compression of the retained air layer with a constant pressure-volume-product $p_{air}V_{air} = p_{amb}V_0$ is assumed, where $V_{air} = V_0 + \int\int F dS$ is the retained air volume, S is the entrapped air surface area and $V_0 = S \cdot h$ is the original retained air volume as shown in Fig. 2.3 b). Let $G(x,y) = F(x,y,0)$ be the critical air-water interface at p_{cr}. The entrapped air pressure can then be expressed as $p_{air} = p_{amb}V_0/V_{air} = p_{air}V_0/\left(V_0 + \int\int G dS\right)$, and Eq. (2.5) can be rearranged to

$$p_{cr} = \gamma \nabla \hat{n}_G - p_{amb}\left[1 - V_0/\left(V_0 + \int\int G dS\right)\right] \tag{2.7}$$

Assuming that the superhydrophobic surface is made up of pores, the air-water interface drooping into a pore takes the shape of a spherical cap as shown in Fig. 2.3 c), and the critical air-water interface can be written as:

$$\begin{aligned} G &= k - o \\ G &= \sqrt{R_{cr} - D^2/4} - \sqrt{R_{cr}^2 - x^2 - y^2} \end{aligned} \tag{2.8}$$

with k being the distance between the origin of the spherical cap and the xy-plane and $o(x,y)$ being the critical air-water interface in respect to the origin of the spherical cap (Fig. 2.3 c). The capillary pressure of a circular pore is defined as

$$p_{cap} = -2\gamma \cos\theta_W / R \tag{2.9}$$

with R being the radius of the pore. Substituting F in Eq. (2.6) by Eq. (2.8) and equalizing the result with Eq. (2.9) gives $R_{cr} = -D/\left(2\cos\theta_W\right)$ as critical radius of the air-water interface, with D being the pore diameter. Finally, inserting the critical radius and Eq. (2.8) into Eq. (2.7) results in an expression for the critical pressure for a surface consisting of pores [73].

$$p_{cr,pore} = \frac{2\gamma}{R_{cr}} - p_{amb} \times \left(1 - V_0 / \left[V_0 + \frac{\pi D^2}{4}\sqrt{R_{cr}^2 - \frac{D^2}{4}} + \frac{2\pi}{3}[(R_{cr}^2 - \frac{D^2}{4})^{3/2} - R_{cr}^3]\right]\right) \tag{2.10}$$

Equation (2.10) predicts that shallow pores exhibit a higher resistance against elevated pressure than deep pores. For shallow pores, the ratio of volume of displaced air to total pore volume is much higher, resulting in higher compression pressures, which in turn stabilize the air-water interface. For deep pores, the critical pressure is almost independent of pore diameter, and only determined by a balance of capillary and hydrostatic pressure, and the contribution of the retained air pressure can be neglected. This can be seen in Eq. (2.10), where the second term becomes obsolete, thus significantly simplifying Eq. (2.10), which then becomes equal to the equation derived by Zheng *et al* [74], where retained air was not considered as stabilizing mechanism. In this case, only the capillary pressure stabilizes the air-water interface.

2.2 Materials

The following section gives an overview over the materials mainly used in this study for fabrication of micro- and nanostructured surfaces as well as peripheral materials not directly structured.

Polycarbonate

Even though nanofur can be fabricated from a variety of polymers [78], this thesis focuses on nanofur fabricated from polycarbonate (PC), namely PC Makrolon LED 2045 from Bayer (Germany). PC is commonly used for everyday products such as DVDs, Blurays or protective goggles, as well as more specialized products such as optical waveguides and other optical instruments [79]. Most PC types are amorphous thermoplasts with a glass transition temperature of $T_g = 144\,^{\circ}\mathrm{C}$ and posses a high mechanical stability [79, 80]. Because light neither gets absorbed nor scattered at its molecules, PC is highly transparent, with transmission values of approximately 89%, negligible absorption and no scattering in the visible and infrared (IR) wavelengths [80]. The excellent optical properties are essential for the intended use as a light managing material and the biocompability of PC further broadens its range of applications. Regarding wetting properties, PC exhibits slightly hydrophilic properties with a static water contact angle of $\theta_W = 72°$ [72, 81].

Cyclo-Olefin-Copolymer

During the fabrication of thin nanofur (see Section 3.1.2) PC is bonded to a cyclo-olefin-copolymer (COC) layer, specifically, COC 6013 F-04 from Topas (Germany). The COC layer serves as a sacrificial protective layer to the thin PC-foils during hot pulling. Similar to PC, COC is a transparent thermoplastic polymer with a glass transition temperature of $T_g = 138\,°C$ [82]. The main criteria for choosing COC was the possibility to bond PC to it without forming a composite, and, thus, being able to separate them from each other after structuring PC [83]. An additional requirement was a T_g in the same range as that of PC, so the polymers could be softened and bonded in one embossing step. Due to the slightly lower glass transition temperature of COC compared to PC, COC softens earlier in the embossing step and encapsulates the edges of the PC-foil, thus, further strengthening the adhesion necessary for a subsequent structuring step of PC.

Fluorinated Ethylene Propylene

Another polymer used for fabrication of microstructured surfaces is fluorinated ethylene propylene (FEP) (grade: FEP500c) from DuPont Corp (USA). Unlike PC and COC, FEP is a semi-crystalline thermoplast with a glass transition temperature of $T_g = 80\,°C$ and a melting temperature (T_m) of $T_m = 260\,°C$ - $280\,°C$ [84, 85]. Depending on its treatment FEP exhibits a degree of crystallisation between 40% and 57% [79]. High transparency, low refractive index ($n = 1.34$) and low surface energy of fluorinated polymers and specifically FEP are the reason why they are often used as top covers for photovoltaic (PV) modules instead of the common glass encapsulation [85, 86]. Especially in the case of flexible and light-weight PV modules, the low weight and flexibility of FEP is superior to the heavy and rigid glass encapsulation. The low surface energy of FEP results in low adhesion of particulate matter on FEP and a water contact angle ($\theta_W \approx 110°$) close to the limitation of unstructured surfaces, which is especially beneficial for self-cleaning properties. Additionally, FEP exhibits high mechanical and chemical resistance, and is not affected by prolonged exposure to ultraviolet (UV) light, which is particularly important for applications where the material is subjected to direct sunlight, such as coatings of PV modules [79, 85].

Polydimethylsiloxane

Polydimethylsiloxane PDMS is used as a protective cover during oxygen plasma etching of thin nanofur and as a soft underground during needle perforation. Needles can puncture samples lying on top of PDMS and penetrate into it without being damaged, as would be the case for hard undergrounds. PDMS is chemically inert, and is not affected by oxygen, which makes it an ideal protective cover during oxygen plasma etching [79]. In this thesis PDMS type Sylgard 184 from Dow Corning (Germany) was used, consisting of a viscous pre-polymer and a hardener. To harden PDMS, the pre-polymer and hardener were mixed in a 10:1 ratio and baked in an oven at 65 °C for 1 hour or left at ambient atmosphere for 24 hours. Because of the viscous nature of the pre-polymer, no limitations exist regarding topography or curvature of surfaces that are to be coated.

2.3 Fabrication Tools and Techniques

Polymer replication techniques such as 3D printing, thermoforming, injection molding or hot embossing allow manufacturing and repeated reshaping of polymers into structured films with structure sizes down to several hundreds of nanometers. Depending on requirements such as complexity, throughput, mechanical properties, material demands and feature size, a suitable fabrication technique has to be found. 3D printing for instances is ideally suited for fast prototyping and fabrication of single units, because it is highly flexible and does not require a costly mold insert. However, even though it becomes ever more versatile in regard to processable materials [87], due to high cycling time and low cost effectiveness, it cannot produce high throughput. In contrast, hot embossing sacrifices the mold free fabrication procedure for higher throughput, and, thus, lower per unit costs. The following section provides an overview over the fabrication and replication techniques utilized in this thesis.

2.3.1 Hot Embossing and Hot Pulling

Hot embossing is a manufacturing process for polymers capable of fabricating structures with feature sizes as small as tens of nanometer [88, 89], and spans a huge variety of materials, geometries and sizes required for different applications [90]. Thermoplastic polymers are the most commonly used material in hot embossing, but the process can be applied to other materials such as bulk metallic glasses [91], ceramics [90], or liquid wood, which is a combination of cellulose fiber and lignin [92]. Hot embossing has already been used as early as 1888 for the replication of vinyl records [90, 93]. While other replication techniques (injection molding, LiGA, thermoforming) have emerged and been refined, the high versatility, low cycling time and, therefore, high throughput, ensure that hot embossing stays a relevant replication technique for small to medium size production batches. Moreover, the possibility of fabricating structures down to a scale of tens of nanometers combined with large fabrication areas of up to 6 inches [94] and the prospect of even greater areas by transferring the process to a roll-to-roll embossing technique ensures its importance in several application fields including microfluidics, lab-on-chip and optical applications [95–98].

In general, hot embossing is comprised of three basic steps as illustrated in Fig. 2.4. First a polymer is placed between the metallic mold insert, which contains the negative of the desired structure, and a suitable counter plate. Even though the mold insert and counterplate do not have to be metallic, choosing a metallic mold

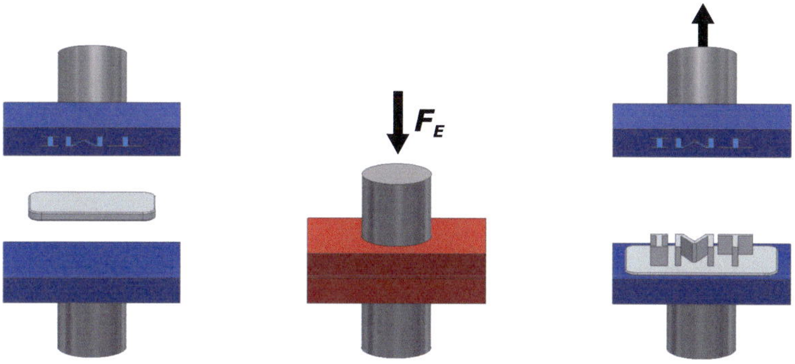

Figure 2.4: Schematic illustration of hot embossing process. A polymer is placed between a structured mold insert (top) and a counter plate (bottom). The setup is then heated above the glass transition temperature of the polymer and the mold is pressed into the melted polymer. Finally, the setup is cooled below the polymer's glass transition temperature and subsequent demolding results in a structured polymer foil.

insert and counter plate ensures fast heat transfer into the polymer as well as high durability of the mold. Next the metallic plates and the polymer are heated above the glass transition temperature of the polymer. The glass transition temperature is defined as the transition temperature between the polymer's energy elastic and entropy elastic state [79, 99]. This transition entails that below the glass transition temperature polymers are in a glassy state and exhibit brittle and stiff properties. An increase in temperature brings the polymer into the glass transisition range, where molecular movement is increased and molecular chains start sliding off each other, resulting in a reduced shear modulus (Fig. 2.5). With temperature increasing above T_g the polymer becomes viscoelastic, its shear modulus decreases further and the polymer enters the flow and melt range, before a further temperature increase results in an irreversible decomposition of the polymer as illustrated in Fig. 2.5. For an ideal hot embossing result, the polymer has to be in the viscosity range highlighted in Fig. 2.5 in order to flow into the cavities without starting to decompose. After reaching the embossing temperature (T_E), an embossing force (F_E) is applied, reducing the gap between mold insert and counterplate

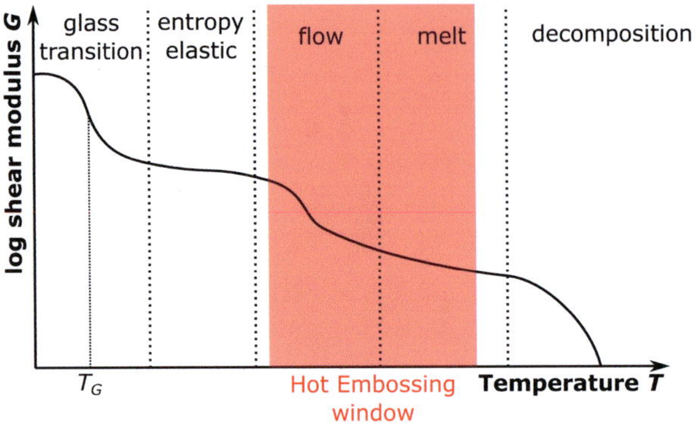

Figure 2.5: Shear modulus of an amorphous polymer as a function of temperature. The embossing window in which the ideal polymer viscosity for high quality hot embossing is reached is highlighted in red. (Adapted from Worgull [90])

and excerting pressure on the polymer melt to press it into the cavities. After waiting for the polymer to completely fill the cavities, the setup is cooled below T_g, where the polymer solidifies in its new form. Finally, the mold insert and counter plate get distanced from one another, removing the structured polymer from the mold insert (Fig. 2.4).

By modifying fabrication parameters in the basic process, replication quality, reproducibility and cycling times can be improved and the replicated structure can be adjusted to application oriented specifications. The most common adjustment is to evacuate the embossing chamber before heating the setup. The removal of air allows the polymer to fill the cavities of the mold insert without encountering resistance. In contrast, under ambient atmosphere, the viscous polymer seals the cavities and traps the remaining air inside, which is then compressed by the flowing polymer and subsequently denies a complete filling of the cavities with polymer, thus, resulting in defects and damages in the replicated structure.

The demolding step can also be adjusted to satisfy specific application oriented requirements. As mentioned above, after cooling down the setup, the structured polymer can be automatically removed from the mold insert through adhesion to the counter plate. To achieve this automated removal, the adhesion between polymer and counterplate needs to exceed the adhesion between mold insert and structured polymer. Hence, the counterplate-polymer adhesion needs to be increased, which is typically achieved by sandblasting the counterplate, and, thus, increasing the counterplate-polymer contact area. The roughness from sandblasting is then transferred onto the polymer backside during embossing, resulting in a highly scattering opaque sample. However, this can often be remedied in a post-processing step if required. Alternatively, manual demolding can circumvent this issue by using a smooth counterplate and removing the polymer from the mold insert by hand after embossing. Because manual demolding does not remove the polymer vertically from the cavities, additional stress is exerted on the polymer topography, which can lead to rupture or breakage of structures. Especially fragile high aspect ratio topographies are prone to get damaged during this process. Another possibility used in this study, is automated demolding with the inclusion of a sacrificial polymer sublayer. In this case, the roughness is transferred to the sublayer and the polymer stack is still vertically demolded, resulting in fully intact transparent samples. However, using a sacrificial polymer requires good adhesion between the sacrificial and structured polymer without forming a composite, to ensure possible separation after demolding. Furthermore, a similar T_g of the polymers is required to ensure adequate adhesion without irreversibly decomposing one of the polymers.

Demolding is also the biggest drawback of hot embossing in regard to reproducible surface geometry, as the removal of the polymer along one axis allows no undercuts relative to this axis. As a result, only topographies with vertical sidewalls or structures rejuvenating towards the top can be replicated, thus, limiting the complexity of the fabricated geometry. Because of this limitation, hot embossing is considered as a 2.5D fabrication technique, able to replicate arbitrary planar geometries, but only able to vary the height of the replicated structures.

Another limiting factor of hot embossing is the achievable aspect ratio (AR) of replicated structures. During demolding, high forces between polymer and mold insert limit the possible AR, making microstructures with AR as high as

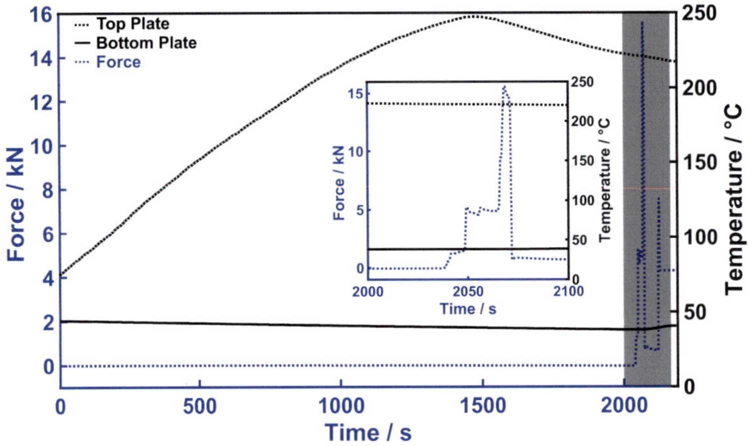

Figure 2.6: Exemplary force and temperature profiles for hot pulling of thin films. No force is exerted until designated embossing temperature is reached. Embossing and demolding take place at a constant high temperature. Inset shows the shaded area in more detail.

10 difficult to replicate [100, 101]. The ratio of surface area to volume increases rapidly for small scale structures, thus, resulting in demolding forces exceeding the polymer cohesion, which will result in the polymeric structure ripping off and clogging the mold insert. To overcome this limitation an adapted hot embossing technique called hot pulling is utilized [49]. Hot pulling differs from the classical hot embossing only in the demolding step. During hot embossing the system is cooled down to ensure a solid polymer before demolding, whereas during hot pulling, the temperature of the mold insert is kept high throughout the complete replication cycle. Figure 2.6 displays experimental force and temperature profiles of an exemplary hot pulling process of a thin film. No force is exerted while the mold insert is heated to the designated embossing temperature. After reaching T_E, mold insert and counter plate are closed, and both embossing and demolding take place at a constant temperature. As a result of the high temperature during demolding, the polymer stays viscous and the high adhesion forces during plate separation are utilized to further elongate the structured polymer, while additionally keeping

the polymer from ripping off and blocking the mold insert. This results in an increased height of the fabricated structures and enables replication of high AR topographies not possible with classical hot embossing.

2.3.2 Plasma Etching

Etching describes the process of removing material by applying chemically agressive substances to a substrate and is divided into dry and wet etching depending on the etchants aggregate state [102, 103]. During this process, an etching mask can be used to protect parts of the substrate surface that are not supposed to be affected by the etchant. In this study polymeric samples are modified using plasma etching, specifically oxygen plasma etching. Oxygen plasma etching is an isotropic dry chemical etching process, resulting in unidirectional removal of exposed material. Etching processes are characterized through the etching rate $r_{etch} = \frac{d_{etch}}{t_{etch}}$, where d_{etch} is the amount of removed material during the applied etching time (t_{etch}). The etching rate is influenced by temperature, sample position, gas pressure and substrate geometry [104]. All etching experiments in this thesis were carried out with a R3T STP2020 oxygen plasma etching machine from MUEGGE GmbH (Germany).

2.4 Microscopic and Surface Analytic Tools

2.4.1 Scanning Electron Microscope

A scanning electron microscope (SEM) is a useful tool to obtain a topographical image of nano- and microstructured surfaces. It works by the same basic principle as a standard light microscope but utilizes electrons instead of light waves for imaging. A beam of electrons is emitted and focused on the sample surface, before being moved over the surface line by line. The highly focused electron beam excites electrons from the surface, which can be collected by a variety of detectors. Depending on the used detector, different information about the surface is gathered. In most cases secondary electrons, which are emitted by surface atoms, are gathered, and, thus, information about the topography of the surface is gained. By using electrons instead of photons, as is the case for a light microscope, a much higher resolution can be achieved and feature sizes ranging from several millimeters down to single nanometers can be resolved. However, electrons can charge non-conductive surfaces, leading to imaging artifacts and a reduced resolution. Therefore, to minimize charging effects and achieve optimal resolution, non-conductive samples are typically coated with conductive metals such as gold or silver prior to imaging, possibly preventing further processing or application of the substrate afterwards. As a result, SEM imaging of coated samples can be classified as a destructive measurement technique for non-conductive substrates. Additionally, similar to a light microscope, images are taken from a top view and can lack crucial information about structure height or hierarchy. However, tilting the substrate or imaging cross-sections, fabricated with sharp razor blades, a microtome or focused-ion beam, enables a complete characterization of a samples topographical features. SEM images throughout this thesis were obtained with a Supra 55P SEM from Zeiss (Germany).

2.4.2 Contact Angle Goniometer

A contact angle goniometer quantifies the wetting behavior of surfaces in regard to a specific liquid. As described in Section 2.1, dispensing a liquid droplet on a surface can result in the droplet spreading completely, staying spherical or displaying any behavior in between. To acquire the CA, the goniometer is equipped with an actuator driven syringe, which is able to dispense drops of various, but well defined, volumes. The dispensed droplet is then illuminated from the backside and the contour is recorded with a camera as either a single image or a movie. A software then analyzes the contour and based on different fitting models calculates the CA (Fig. 2.7). The three different fitting models used in this thesis are ellipse fitting, Young-Laplace fitting and manual fitting and are later described in more detail.

The goniometer can further be used to measure the advancing and receding contact angle on surfaces. For this, a droplet is dispensed onto the surface while still attached to the needle. Then, the volume of the droplet is increased and subsequently decreased and a movie of the evolving droplet shape is recorded. The advancing contact angle is measured as the highest CA during volume increase and the receding contact angle is measured as the lowest CA during deflation of the droplet. Combining the sample holder with a tilting unit further allows measurement of the sliding angle and contact angle hysteresis. After dispensing a droplet on the surface and retracting the needle, the substrate is tilted until the droplet starts to slide off the surface. The maximum tilt angle at which the droplet is still attached to the surface describes the sliding angle, while the angle at the front/rear side of the droplet before sliding off is $\theta_{adv}/\theta_{rec}$.
Static contact angles in this thesis were measured with the OCA 15Pro contact angle measurement system from DataPhysics Instruments. To record sliding angles the measurement system was rotated using an electronic tilting base unit (TBU90E, Dataphysics, Germany).

Fitting Types and Errors

As mentioned above, the CA is calculated from a recorded drop shape profile by curve-fitting software using different fitting methods. Depending on the contact angle, shape and homogeneity of the substrate, various fitting errors can occur and a suitable calculation method has to be chosen to achieve the best result (Fig. 2.7).
Young-Laplace fitting considers the whole droplet shape and assumes gravity is the only force acting on the droplet. Furthermore, for this fitting type to be applicable, the droplet needs to exhibit a high degree of symmetry, which is the reason why it is also called axisymmetric drop shape analysis (ADSA) [105]. Using the surface tension of the liquid, the ADSA method iteratively solves the Laplace equation, and, thus, calculates the CA. In contrast to all other fitting methods, it takes physical properties of the droplet into account and is not only based on geometrical curve fitting. Especially for high CA the ADSA method

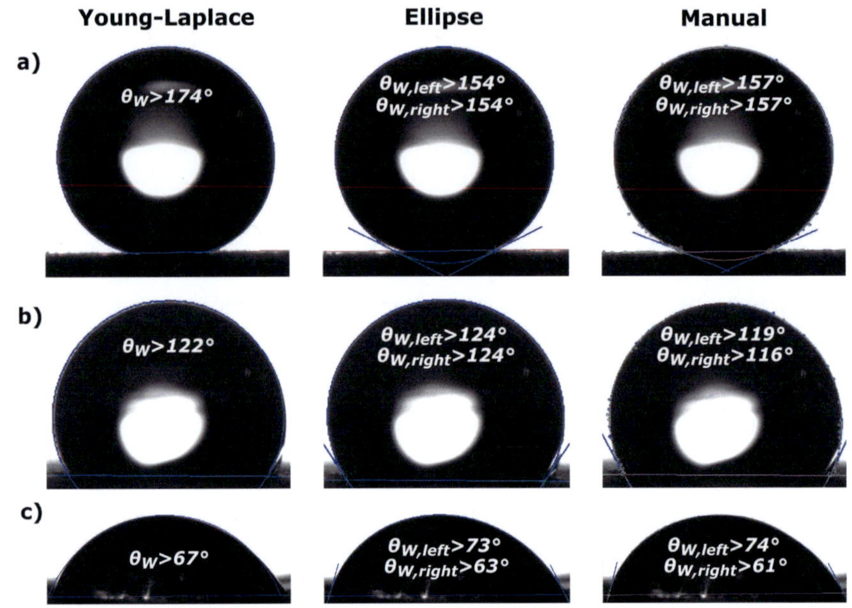

Figure 2.7: Contact angle fitting types and their failures. The contour of three droplets are shown with the CA calculated based on Young-Laplace fitting (first column), ellipse fitting (second colum) and manual fitting (third column) showing the respective errors. a) For high contact angles, Young-Laplace shows the best fit, while ellipse and manual fitting underestimate the CA. b) Optical defects prohibit automatic fittings (Young-Laplace, ellipse) from tracing the droplet contour. c) Assymetric droplets cannot be traced by Young-Laplace fitting, while both ellipse and manual fitting are capable of distinguishing directional CA.

traces the droplet profile best and shows the most accurate and stable fitting, while ellipse and manual fitting underestimate the CA, as shown in Fig. 2.7 a).

Ellipse fitting is a geometrically based fitting mode. It assumes the droplet to have an elliptical shape and fits an ellipse equation to the captured droplet. The CA is then calculated as the angle of the tangent in the contact points of the ellipse and the baseline. Ellipse fitting does not assume a symmetric shape and, therefore, can calculate separate CAs for the left and right side of the droplet. As a result, it is a good fit for asymmetrical droplets and heterogeneous surfaces, where Young-Laplace fitting cannot be used due to its assumption of symmetry (Fig. 2.7 c). However, for high CA the assumed elliptical shape cannot accurately trace the droplet profile and, therefore, ellipse fitting typically underestimates the CA.

Manual fitting is a subroutine of ellipse fitting and, therefore, also a geometrically based fitting mode. For manual fitting at least five spots of the droplet profile need to be manually defined. Then an elliptical equation is fitted to the manually defined points and interactively adjusted to each newly defined profile point. Similar to ellipse fitting, the CA is then calculated as the tangent of the elliptical equation in the contact points of the ellipse and the baseline. Manual fitting is especially beneficial when distortions due to light make an automatic fitting of the captured droplet impossible, as shown in the lower right part of Fig. 2.7 b).

Circle and tangent fitting are other commonly used fitting routines. Circle fitting uses the same algorithm as ellipse fitting, with the exception of fitting a circle to the detected droplet. Tangent fitting extracts a set amount of points of the droplet contour above the baseline and numerically calculates a tangent based on the extracted data. Both routines have shown less accurate results than Young-Laplace, ellipse and manual fitting, and are therefore not further considered in this work.

Because most investigated surfaces in this thesis were superhydrophobic with $\theta_W > 150°$, Young-Laplace fitting was used for all static contact angle measurements unless otherwise specified.

2.4.3 Wilhelmy Plate Tensiometer/Contact Line Motion

Characterization of the dynamic wetting behavior of surfaces being immersed into water is crucial for many applications and complimentary to the wetting behavior measured with a contact angle goniometer. A Willhelmy plate tensiometer is a useful tool to measure the dynamic contact angles of surfaces and investigate properties not accessible otherwise [106, 107]. Typically a Wilhelmy plate tensiometer is used to measure the surface tension of liquids [108]. It consists of a plate of length l and width d, which is attached to a microbalance and immersed

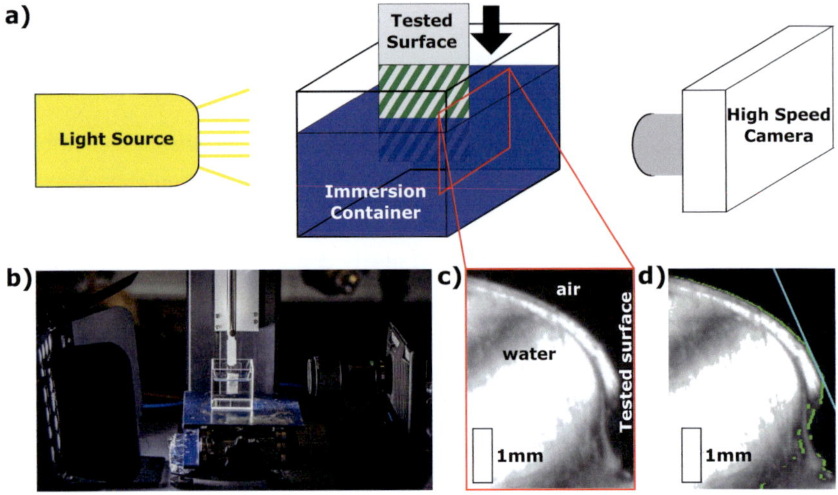

Figure 2.8: Contact line motion measurement. Schematic (a) and photograph (b) of contact line motion setup. A substrate is illuminated from the side, while being immersed in a test liquid. A high speed camera records the evolution and movement of the contact line. c) Exemplary image of the recorded contact line on a superhydrophobic thin nanofur. d) Air-water interface and tangent are extracted and highlighted for CA calculation.

in a testing liquid. The microbalance measures the triple-line anchoring force (F) and by using $\gamma = F/(u\cos\theta_W)$, where $u = (2w+2d)$ is the wetted perimeter, makes the calculation of the surface tension (γ) possible [108].

Dynamic CAs are measured by observing the contact line motion on an adapted Wilhelmy plate tensiometer as shown in Fig. 2.8. Superhydrophobic surfaces are attached to a glass slide and by using a dip coater immersed in a testing liquid at a controlled immersion velocity. During immersion, a light source illuminates the sample from the side, while a high speed camera records the evolution and movement of the contact line (air-water interface). The contact line is then extracted through a custom Matlab code (Appendix) and highlighted. Using the extracted contact line, the Matlab code then semi-automatically fits a tangent to the air-water interface by linear regression of the liquid contour before the solid-liquid-air triple point, and calculates the dynamic contact angle as the angle between the

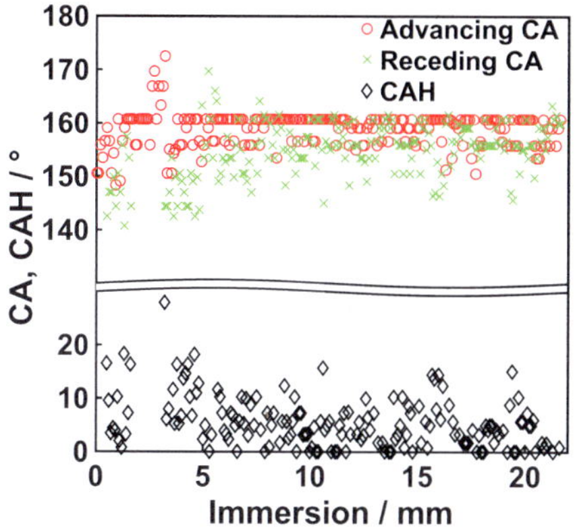

Figure 2.9: Position resolved dynamic wetting behavior. Locally resolved advancing and receding CA and CAH as a function of immersion depth for a thin nanofur film immersed in deionized (DI) water at 20 mm min^{-1}.

tangent and the vertical substrate. In contrast to sessile drop measurements with the goniometer, which only measure the wetting properties on individual localized spots, the semi-automatic CA calculation allows a fast characterization of the dynamic wetting properties of the complete surface (Fig. 2.9).

Furthermore, the measurement technique allows a characterization of surfaces while being immersed and moved through a liquid and allows an allocation of wetting behavior (CA and CAH) to a specific location on the substrate as shown in Fig. 2.9, thus, mapping the sample. A constant, in-situ CA measurement is capable of showing Cassie-Baxter to Wenzel wetting transitions as soon as they occur as a plummeting dynamic contact angle, and, thus, is able to measure the stability of the superhydrophobic state, while the surface is in motion. Furthermore, mapping of the sample through precise localized contact angle measurements allows correlation of dynamic wetting behavior and wetting transitions to such properties as roughness, surface topography and slip length accessible through other locally resolved measurement techniques, such as atomic force microscopy, SEM or vertical scanning interferometry.

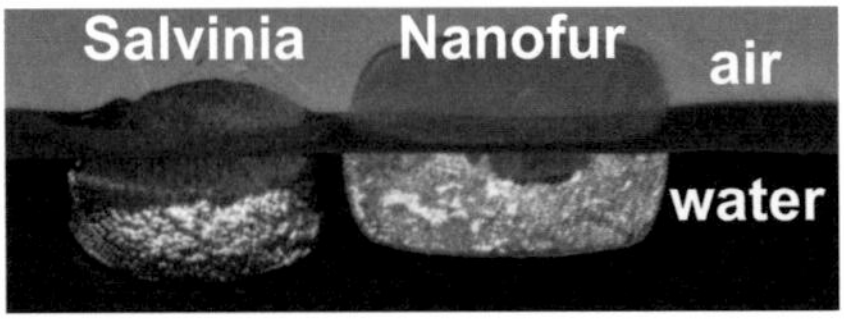

Figure 2.10: Retained air layer on superhydrophobic surfaces. Photograph of nanofur and *Salvinia molesta* leaf submerged underwater. Silver shimmer is caused by total internal reflection at air-water interface and indicates the retained air layer. (Adapted with permission from [81]. Copyright 2015 American Chemical Society.)

Moreover, by varying immersion liquid, immersion velocity, liquid temperature, and, thus, capillary number (Ca[1]) and Reynolds number[2], as well as immersion angle, substrate temperature, ambient temperature, ambient pressure and surrounding gas/liquid, the measurement setup can easily be tailored to represent specific application oriented environments.

2.4.4 Underwater Retained Air Layer Measurement

When superhydrophobic surfaces are submerged under water they retain an air layer which can be seen as a silvery shimmer due to total internal reflection (TIR) at the air-water interface (Fig. 2.10). When a wetting transition occurs, the sample looses its superhydrophobicity and water infiltrates the topography and replaces the retained air layer. The breakdown of the retained air layer results in a visible loss of reflectivity, best observed from a tilt angle (α) close to the critical tilt angle $\alpha_{\mathrm{TIR}} = \arcsin(n_2/n_1)$, with n_1/n_2 being the refractive index of air/water ($n_1 = 1.0 / n_2 = 1.33$). The high reflectivity and change thereof during wetting transitions can be utilized to measure the stability and evolution of the retained air layer. To do this, superhydrophobic surfaces are submerged in a container filled with deionized (DI) water and optical images of the air-water interface are recorded with a camera tilted at α_{TIR}, while the sample is exposed to external

1 $\mathrm{Ca} = \mu v/\gamma$, describes the relation between viscous forces and surface tension, with μ being dynamic viscosity and v characteristic velocity.

2 Reynolds number is a dimensionless number predicting flow patterns. A low/high Reynolds number typically describes a laminar/turbulent flow.

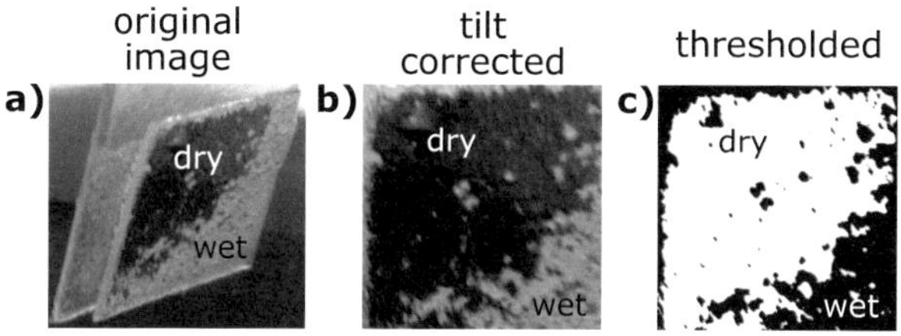

Figure 2.11: Extraction of retained air layer area. a) The intact retained air layer on a submerged superhydrophobic surface reflects the black background, while the wetted area is white. b) Photograph is tilt corrected and c) thresholded to extract the dry area.

stimuli such as pressure or movement. To increase visibility and contrast of the retained air layer compared to the wetted area, a black background is installed next to the container, making the air-water interface appear black (Fig. 2.11 a). A wetting transition (collapse of the retained air layer), and the associated loss of reflectivity, results in a change from black (dry area) to the original white sample colour (wet area). To extract the dry area of the sample, recorded images are tilt corrected, converted to greyscale and thresholded as shown in Fig. 2.11. The dry area of each sample is then normalized to the dry area before the superhydrophobic surface is exposed to external stimuli, to allow direct comparison between different samples and other superhydrophobic surfaces.

2.4.5 Confocal Laser Scanning Microscopy

The shape and position of an air-water interface spanning over a superhydrophobic surface can be visualized using a confocal laser scanning microscope (CLSM) [50]. A confocal laser scanning microscope (CLSM) spatially records the reflection intensity of an illuminated spot in the focal plane, while ignoring most of the out-of-focus reflection from areas above or below the focal plane. As a result, the CLSM is very sensitive to changes of refractive index at material interfaces, as these increase reflectivity, and, therefore, appear bright in the recorded image (Fig. 2.12 b). Inside a medium reflection is minimal and diffusion is dominant, resulting in less light being recorded, and, as a result, these areas appear as black

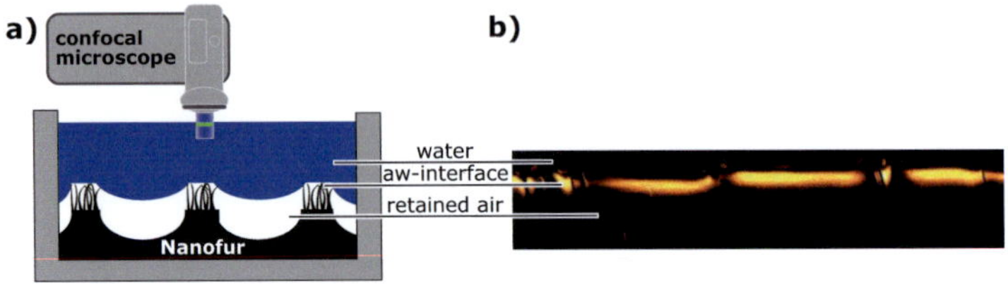

Figure 2.12: Visualization of the air-water interface. a) Schematic illustration of the CLSM measurement setup. b) Cross-sectional CLSM image of air-water interface (aw-interface) spanning submerged superhydrophobic thin nanofur. Bright areas indicate air-water interface separating water (dark area above interface) from retained air layer (dark area below interface).

regions in the image. An automated shift of the focal plane enables the recording of vertical image series of a sample, and, thus, cross-sectional visualization of the sample and its interfaces (Fig. 2.12 b). To visualize the air-water interface of a superhydrophobic surface, the sample is immersed in a container filled with DI water and the CLSM records a vertical series of the sample and the retained air layer (Fig. 2.12 a). An exemplary air-water interface spanning a thin nanofur film is shown in Fig. 2.12 b), where the bright areas indicate the jump in refractive index at the air-water interface, separating the water layer (dark area above the interface) from the retained air layer (dark area below the interface). Combining the CLSM with a pressure cell enables visualization of the effect hydrostatic pressure has on the shape of the air-water interface suspended over a superhydrophobic surface (Section 4.1.1).
Cross-sectional images of air-water interfaces in this thesis were recorded using a Leica TCS SP2 CLSM from Leica Mikrosysteme GmbH (Germany).

2.4.6 Spectrophotometer

The macroscopic interaction of light with matter in this thesis is measured using a LAMBDA 1050 UV/VIS/NIR spectrophotometer from PerkinElmer (USA). Calibrated UV, VIS (visible) and NIR (near infrared) light sources are capable of emitting light in the range of 175 nm to 3300 nm. A highly reflective, white integrating sphere is used to measure the total (diffusive and specular) transmission and reflection of samples using the setup schematically depicted in Fig. 2.13. To measure total transmission, the sample is positioned in a sample holder at the entrance of the integrating sphere, while total reflection can be measured by placing the sample at the backside of the sphere tilted at 8°. Moreover, the sample can be attached to a rotatable sample holder, which can be placed in the middle of the integrating sphere, in order to measure angle of incidence (AOI) resolved reflectance. Equipping the spectrophotometer with a three detector module allows measurement of the specular transmission, by placing the sample 9 cm away from the entrance to the detector.

2.4.7 External Quantum Efficiency Measurement

External quantum efficiency (EQE) describes the ratio of charge carriers collected by a photovoltaic (PV) device to the number of photons impacting on a PV device (photon-to-current efficiency) and is usually given as a function of wavelength [109]. Hence, measuring EQE is a useful tool to characterize efficiency of PV devices and can be used to show the effect an attached optical coating has on the

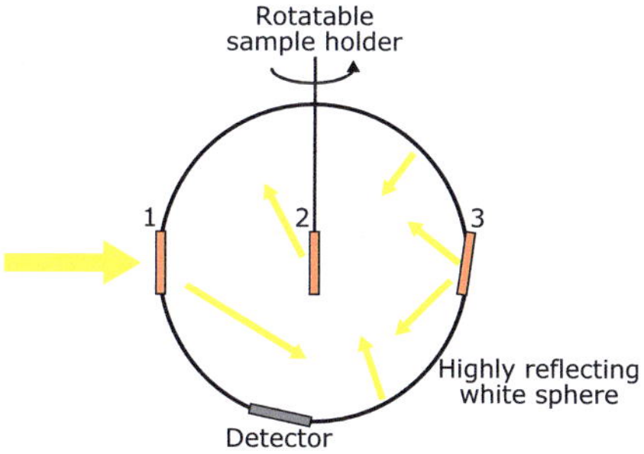

Figure 2.13: Schematic illustration of spectrophotometric measurements. Monochromatic light is guided towards the integrating sphere and depending on the sample position reflection or transmittance is measured. Total transmission is measured if the sample is mounted at the detector entrance (position (1)). To measure total reflection, the sample is mounted at the backside of the integrating sphere tilted at 8° (position (3)). For angle resolved reflection measurements, the sample is mounted on a rotatable sample holder in the center of the integrating sphere (position (2)).

electrical output of PV modules. During the EQE measurement the tested solar cell is typically in the short-circuit condition and illuminated with monochromatic light, while a white light "bias", simulating sunlight (air-mass 1.5 global spectrum [110]) is added. In this thesis, the EQE of multicrystalline silicon (mc-Si) solar cells from E-Ton Solar (Taiwan) was measured using a xenon lamp and a monochromator-based spectral response system from Oerlikon Solar (Switzerland). One-sun bias light was provided by a halogen lamp. Efficiency enhancing films were optically coupled to the solar celly by dispensing several droplets of refractive matching liquid on the films backside, before pressing the film onto the solar cell with enough physical force to prohibit formation of air bubbles between the sample and the solar cell. For multicrystalline solar cells several spots need to be measured and averaged, to account for the efficiency discrepancies due to the multicrystalline nature of the cell.
Another characteristic parameter for solar cells is the short-circuit current density (J_{SC}), which can be calculated from the EQE and the incident spectrum (air-mass 1.5 global spectrum [110]) using the following equation.

$$J_{SC} = q \int_{\lambda_1}^{\lambda_2} EQE(\lambda)\Phi(\lambda)d\lambda \tag{2.11}$$

Here q is the elementary charge, $\Phi(\lambda)$ is the incident spectrum's photon flux and $\lambda_1 = 300\,\text{nm}$ and $\lambda_2 = 1100\,\text{nm}$ define the relevant spectrum. The short-circuit current density describes the theoretical maximum current that can be drawn from a solar cell divided by the area of the solar cell. Therefore, the higher the J_{SC}, the higher the electrical output of a PV module.

3 Fabrication and Wetting Properties of Nanofur

Inspired by natural surfaces scientists have been exploring new ways and methods to fabricate superhydrophobic surfaces for a long time [11, 42, 71, 111–113]. However, developing a scalable and cost-effective technique to fabricate superhydrophobic surfaces remains challenging to this date.
In this chapter I will introduce upscalable fabrication procedures for bioinspired superhydrophobic surfaces based on hot embossing and hot pulling. First, in Section 3.1.1, the state of the art fabrication procedure of nanofur, developed by Röhrig *et al.* [49–51], will be presented. Then, an adapted fabrication technique reducing thickness and increasing flexibility of the nanofur is introduced in Section 3.1.2, resulting in "thin nanofur". Section 3.2 introduces the fabrication of a fluorinated superhydrophobic microcavity array (MCA) based on nanofur topography and used to enhance the electrical output of PV modules, while Section 3.3 describes the fabrication of two other superhydrophobic surfaces (double scale teflon surface wrinkles (DS wrinkles) and polytetrafluorethylene (PTFE) pillars) to which the retained air layer stability of thin nanofur is compared to later in the thesis. Finally, in Section 3.4 the wetting properties of thin nanofur and fluorinated ethylene propylene (FEP) MCA are investigated.

3.1 Nanofur Fabrication

3.1.1 Thick Nanofur Fabrication

Superhydrophobicity of water fern *Salvinia* and water lettuce *Pistia stratiotes* originate from a dense layer of microhairs covering their surface [11, 70, 114].

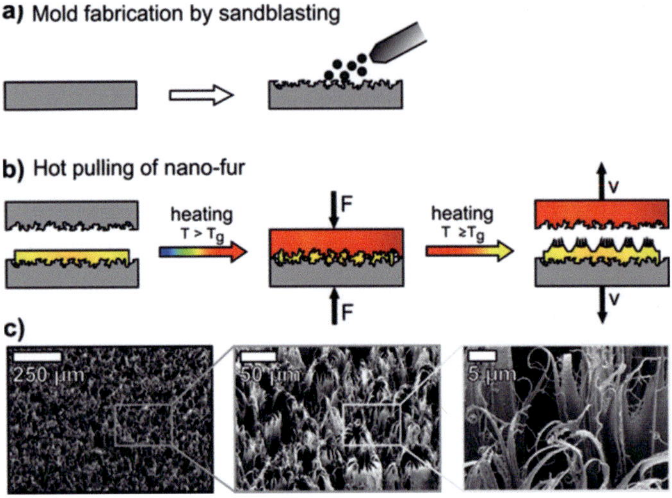

Figure 3.1: Fabrication of thick nanofur by hot pulling. a) A sandblasted steel-plate serves as mold insert. b) The heated mold is pressed into the foil in an automated process. The polymer softens and fills the cavities of the mold. By retracting the heated mold, the softened polymer elongates and forms a fractal surface covered with high aspect ratio nanohairs. The foil adheres to the lower plate due to its high roughness which is also produced by sandblasting. c) SEM micrographs showing the fractal surface topography of nanofur fabricated from polycarbonate. Depending on the processing parameters density, length, and width of the hairs can be varied in a wide range. (Reprinted with permission of Röhrig *et al.* [50]. Copyright 1999-2018 John Wiley & Sons, Inc.)

Inspired by these natural prototypes Röhrig *et al.* [49–51] used a hot pulling technique to fabricate a polymeric nanofur film covered in a dense layer of nano- and microhairs. The fabrication process is schematically shown in Fig. 3.1. First a steel plate is sandblasted with aluminum oxide (particle size: $23 \pm 1.5\,\mu m$) in order to be used as mold insert during the fabrication process. Then, after evacuating the hot embossing chamber, the mold insert is heated above the glass transition temperature of PC ($T_g = 144\,^{\circ}C$), after which, the sandblasted steel plate is moved towards an approximately 2 cm x 2 cm large, 1 mm thick PC-plate fixated to another sandblasted (silicate clinker, particle size: 0.6 - 1.5 mm) steel plate. Upon contact, the polymer softens and flows into the roughness of the mold insert. The mold is then pressed into the polymer with an embossing force of

$F_E = 20$ kN, before being retracted while still being hot. Adhesion to the top plate and high demolding forces during retraction elongate the softened polymer, resulting in a layer of microcavities surrounded by high aspect ratio micro- and nanohairs (Fig. 3.1 a). The so fabricated microhairs can be up to several microns long and display tip diameters of less than 200 nm [50].

3.1.2 Thin Nanofur Fabrication

Because this thesis focuses on the application of superhydrophobic nanofur in optoelectronic devices and air retaining applications, the fabricated nanofur needed to exhibit a minimal thickness and high flexibility, in order to increase transparency and enable further surface processing. Therefore, the fabrication process was adapted to fabricate so-called "thin nanofur", with a more than 30% reduced thickness compared to previously fabricated "thick nanofur". An additional hot embossing step was included prior to hot pulling of the nano- and microhairs, and the adapted fabrication procedure is schematically shown in Fig. 3.2 a), b). Similarly to thick nanofur fabrication sandblasted steel plates are used as mold insert and counterplate. To fabricate thin nanofur films, 2 cm x 2 cm large optical grade PC-foils with thickness ranging from 100 µm to 200 µm were used as base material. However, because polymer fixation to the bottom plate and sufficient material for structuring are prerequisites for hot pulling from a thin film, thin PC-foils were first bonded to a sacrificial COC layer (Fig. 3.2 a). COC was chosen as sacrificial polymer as well as an embossing temperature of $T_E = 160\,^{\circ}$C and an embossing force of $F_E = 2$ kN to ensure possible separation of thin PC nanofur from the sacrificial layer, after structuring it by hot pulling. These parameters ensure good adhesion between PC and the sacrificial COC-layer without forming a composite [83]. Furthermore, using a sacrificial polymer layer during fabrication has the additional advantage that roughness from the bottom sandblasted steel plate is transferred onto the sacrificial layer instead of the backside of the thin nanofur, as is the case for the original "thick nanofur". This reduces reflection losses, which is essential for optoeletronic applications described in Chapter 5. After fixation

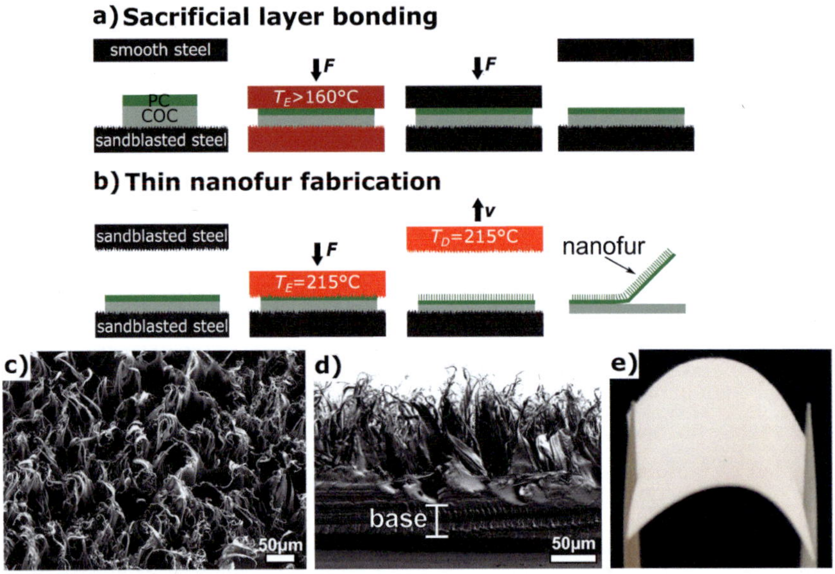

Figure 3.2: Fabrication of thin nanofur. a), b) Schematic of the fabrication process. First PC and COC are bonded together in a hot embossing step. Then the heated top mold is pressed into the polymer stack and the softened PC creeps into the cavities of the finely sandblasted top mold. Retraction of the hot mold elongates the softened PC and creates microcavities surrounded by a dense layer of nano- and microhairs. The thin nanofur film is then separated from the sacrificial COC layer. c) SEM image of thin nanofur. d) Cross-sectional SEM image of thin nanofur film with an average hair length of 110 µm and a minimal base thickness of 35 µm. Minimal hair tip diameter is below 200 nm. e) Photograph of highly flexible thin nanofur film. (a) - d) reprinted with permission from Vüllers *et al.* [72]. Copyright 1999-2018 John Wiley & Sons, Inc.)

of the polymer stack to the bottom plate, high aspect ratio nano- and microhairs are formed during a hot pulling step as shown in Fig. 3.2 b). Similar to thick nanofur fabrication, a finely sandblasted steel plate is heated to 215 °C (well above T_g of PC), before being pushed 200 µm into the polymer stack at 10 mm min^{-1}, thus ensuring that the thin PC-foil does not get punctured. During this step, an embossing force of approximately $F_E = 6.5$ kN is exerted onto the polymer stack. As soon as the mold insert contacts the polymer, it softens and flows into the cavities of the mold. After reaching the defined embossing depth, the top mold

Figure 3.3: Surface topography of *Salvinia cucullata* and *Pistia stratiotes*. Cross-sectional microscopy images of a) *Salvinia cucullata* and b) *Pistia stratiotes* leaves covered with transparent microhairs. Insets show photographs of the superhydrophobic leaves of a) *Salvinia cucullata* (Image: W. Barthlott, University of Bonn) and c) *Pistia stratiotes*. (Figure adapted with permission from Vüllers *et al.* [72]. Copyright 1999-2018 John Wiley & Sons, Inc.)

is retracted at 100 mm min^{-1} and with a demolding temperature $T_D = 215\,°C$, thus, resulting in a cratered surface covered with high aspect ratio micro- and nanohairs (Fig. 3.2 c). The dense layer of hairs is up to 150 µm high with hair tip diameters down to 200 nm and is supported by a thin residual PC-layer with thickness ranging from 30 µm to 150 µm as shown in Fig. 3.2 d). Overall thickness of thin nanofur films can be varied in a range of 70 µm to 300 µm by modification of original PC foil thickness and fabrication parameters. Low film thickness makes the thin nanofur film highly flexible as shown in Fig. 3.2 e). Comparing the surface topography of both thick and thin nanofur to their biological inspiration *Salvinia cucullata* and *Pistia stratiotes* (Fig. 3.3) reveals many important similarities, as all of the surfaces are covered in dense layer of microhairs. These topographical similarities also manifest themselves in the excellent wetting and optical properties of thin nanofur, as discussed later in this thesis (Section 3.4.1, Section 5.1).

Variation of hair density on thin nanofur

Röhrig *et al.* [50] have shown that by varying fabrication parameters, both hair density and diameter of microcavities of thick nanofur can be influenced, thus, even changing the wetting behavior of nanofur. To investigate the effect hair density has on the dynamic stability of retained air layer in Section 4.2, superhydrophobic thin nanofur films with varying hair densities were fabricated. While high density (HD) thin nanofur is fabricated as described above, a variation of the embossing temperature by 8 °C during hot pulling can drastically change the hair density and result in medium density (MD) and low density (LD) thin nanofur (Fig. 3.4). All three nanofur topographies display similar superhydrophobic properties and are able to retain an air layer when submerged underwater.

It has to be noted, that the low complexity of mold fabrication by sandblasting a steel plate makes upscaling of nanofur fabrication to a standard roll-to-roll polymer processing technique feasible [50, 90]. Furthermore, due to high forces during the demolding process no other conventional replication techniques are able to fabricate such high aspect ratio micro- and nanostructures without destroying the mold insert [115, 116]. In this thesis, unless otherwise specified, nanofur is used synonymously with thin HD nanofur, i.e. nanofur made from thin optical grade PC.

3.1.3 Perforated Nanofur Fabrication

Chapter 4.1 focuses on increasing the stability of the retained air layer against external stimuli such as hydrostatic pressure. In order to do this, access to the retained air layer of nanofur submerged underwater has to be gained. This is achieved by perforating nanofur with microneedles, plasma etching and laser drilling as described in the following. For all three perforation techniques to be successful, a minimal nanofur thickness, and, thus, thin nanofur fabrication is essential.

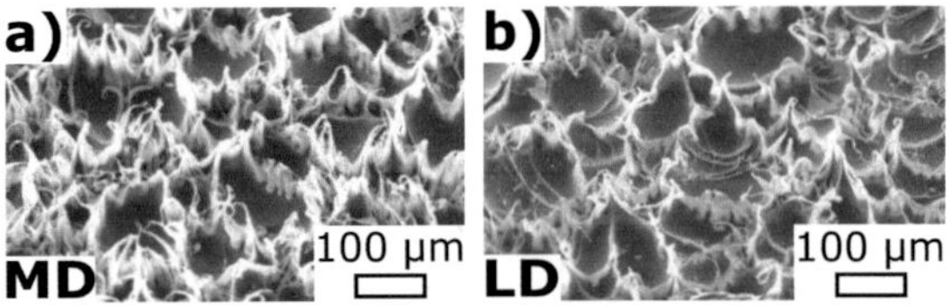

Figure 3.4: Thin nanofur with varying hair density. Variation of fabrication parameters decreases hair density and increases cavity diameter. SEM image of medium density (MD) (a) and low density (LD) (b) nanofur.

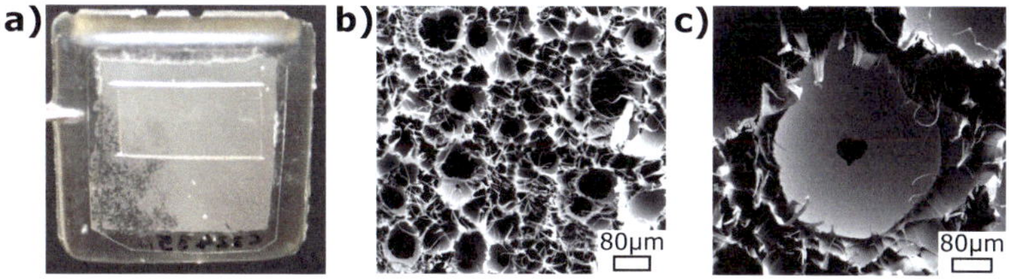

Figure 3.5: Oxygen plasma etching of nanofur. a) Photograph of PDMS encapsulated thin nanofur with cut out etching window. b), c) SEM images of thin nanofur after oxygen plasma etching. Etching process opened cavities resulting in pores with diameters ranging from 15 µm up to 150 µm, while not damaging surrounding micro- and nanohairs.

Perforation by Plasma Etching

One possible perforation technique is oxygen plasma etching. The objective of this technique was to etch nanofur from the backside, thus, removing material from the residual layer and opening up the cavities at the lowest point without harming the hairs, because removal of hairs would result in a loss of superhydrophobicity. PC nanofur was etched at 1200 W and 22 °C resulting in an etching rate of 2 µm min^{-1}. For the first etching test, thin nanofur was simply placed upside down in the etching chamber, thus, having its backside face the plasma source. Etching with this configuration opened the pores as desired, but also damaged the micro- and nanohairs during the etching process.

Therefore, to protect the nanofur topography during the etching process, the film was encapsulated in a protective PDMS layer, acting as an etching mask. For encapsulation, thin nanofur was covered with a 10 : 1 mixture of PDMS and curing agent and cured at 65 °C in an evacuated oven. To be able to etch pores into the thin nanofur, a small window of PDMS is removed from the backside of the nanofur revealing an etching window as shown in Fig. 3.5 a). By removing only part of the PDMS coating, the actively etched area can be precisely controlled.

Depending on nanofur thickness, 65 µm to 80 µm PC was etched away, before the rest of the protective PDMS layer was peeled off thin nanofur without damaging the nanofur topography as shown in Fig. 3.5 b), c). The SEM-images show that etching succeeded in opening cavities, and, thus, perforating the nanofur without harming the microhairs. However, the pore diameter ranged from 15 µm up to 150 µm on one sample. Additionally the variation of residual thickness and differences in localized etching rate, can result in every cavity opened. With every cavity opened, the overall structural integrity of the thin nanofur film is reduced, and handling of these excessively perforated thin nanofur films or exposure to minimal mechanical stress can easily rupture and irreversible damage the films. Because of these issues and the limited control over the pore diameter, perforation by etching was not further pursued as a perforation technique for thin nanofur.

Perforation by Laser Drilling

Another perforation technique investigated in this thesis is laser drilling. Laser are typically used in material fabrication of thin films because of the minimal thermal and non-existent mechanical and electrical stress during fabrication [117]. In order to minimize damage to the nanofur topography a short pulse laser operating at a wavelength of 248 nm, a frequency of 200 Hz and a pulse length between 4 ns and 6 ns was used. Laser perforation was undertaken by Heino Besser from the Laser Process Technique Group at the Institute of Applied Materials. Thin nanofur films were laser drilled from the backside, as drilling from the front side damaged the nano- and microhairs, and, thus, resulted in a loss of superhydrophobicity. In contrast, laser drilling from the back did not damage the nanofur topography and resulted in pore diameters down to 15 µm (Fig. 3.6). Additionally, pores are drilled according to a predefined array, and, thus, position and number of pores can be controlled precisely. Because laser-drilling can be completely automated, it is a feasible method for both perforation of large scale films as well as perforation of a high number of films.

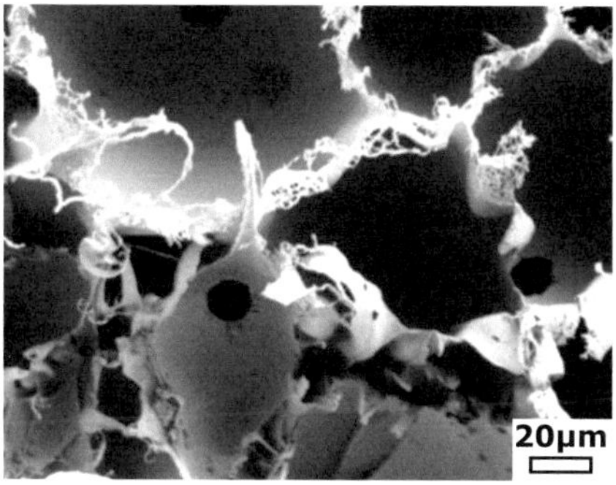

Figure 3.6: Laser drilling of thin nanofur. SEM image of thin nanofur film displaying laser drilled pores with a diameter of approximately 15 μm.

Perforation by Microneedles

The third perforation technique employed is manual perforation of thin nanofur using microneedles with a nominal diameter of 100 μm. Thin nanofur was placed on a soft supporting layer of PDMS and then perforated by hand with a microneedle, resulting in pores with diameters ranging from 90 μm to 120 μm [118]. Similar to previous perforation techniques, manual perforation only removes the material inside the pore, without damaging the surrounding surface topography of micro- and nanohairs as shown in Fig. 3.7.

Manual perforation of thin nanofur gives precise control over positioning and number of pores. Moreover, the pore diameter can be changed by changing the diameter of the needle and is only limited by availability of respective needles. One drawback of manual perforation is the high amount of time needed for fabrication of every single pore, making manual perforation a time-consuming modification technique. Additionally, needles with small diameter can be fragile and easily broken if not inserted vertically into the sample. However, using a stamp equipped with multiple microneedles or a microroller makes a large scale perforation possible, and offers additional support, which increases stability and durability of the microneedles [118, 119]. Unless otherwise specified, perforated nanofur always describes manually perforated nanofur.

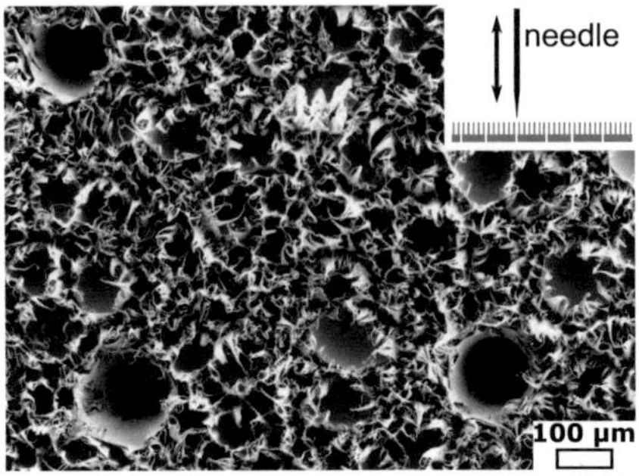

Figure 3.7: Microneedle perforated nanofur. SEM image of thin nanofur film perforated with a microneedle with diameter 100 µm.

3.2 Fabrication of FEP Microcavity Arrays

Numerical simulations and experimental results in Section 5.2 show that a microcavity array (MCA), as found on thin nanofur supporting the dense layer of hairs, is ideal for optical applications. However, to achieve superhydrophobicity and self-cleaning properties with low aspect ratio MCA structures, the structure has to be imparted to a polymer with low surface energy, such as a fluorinated polymer. In order to transfer the disordered MCA topography from thin nanofur to FEP, the fabrication process displayed in Fig. 3.8 a) was developed. First a nickel shim was created by galvanically templating a PC nanofur film with the help of Dr. Markus Guttmann (IMT, KIT). To do this, thin nanofur was attached to a silicon wafer, before 8 nm chromium (adhesive layer) and 150 nm gold (conductive layer) was evaporated onto the polymer. After electrically connecting the polymer substrate and the wafer with copper tape, the nanofur substrate was masked with several layers of adhesive green tape creating a 1 mm thick and 22.5 mm x 22.5 mm large plating window. Then, the masked wafer was immersed in a nickel electroplating system with standard boric acid containing nickel sulphamate electrolyte ($T = 52\,°C$, pH ranging from 3.4 to 3.6), specifically designed for electroforming micro- and nanostructures [121, 122]. By ramping the current density from $1 \cdot 10^{-3}\,A\,cm^{-2}$ to $1 \cdot 10^{-2}\,A\,cm^{-2}$ throughout the plating process, a defect-free 600 µm thick Ni-layer was electroformed consisting of densely packed microhemispheres (Fig. 3.8 a), b)). Finally, the silicone wafer was removed from the Ni-shim through wet chemical dissolving and residual polymer was cleaned off the shim by exposing it to oxygen plasma for 150 min at 22 °C and 1200 W.

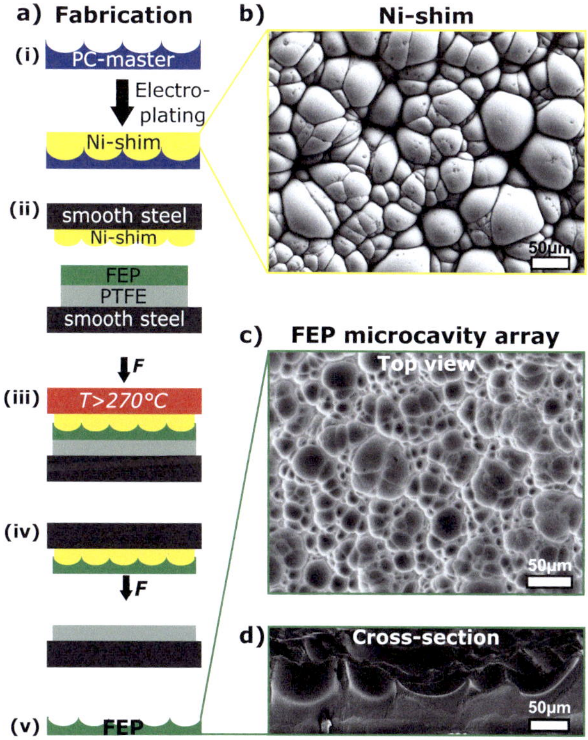

Figure 3.8: Fabrication of fluorinated ethylene propylene (FEP) microcavity array (MCA). a) Schematic of Ni-shim fabrication and hot embossing process. The MCA on PC nanofur is galvanically copied by electroforming to fabricate a Ni-shim. The shim is then attached to a steel plate, heated and pressed into a FEP foil resting on a PTFE plate, thus, embossing the MCA topography into FEP. After cooling, the Ni-shim is removed from the FEP, resulting in a FEP MCA. b) SEM image of the microstructured surface of the Ni-shim. c) SEM image of the embossed FEP MCA surface topography. d) Cross-sectional SEM image of the FEP MCA reveals inverted hemispherical indentations in FEP film. (Reprinted with permission from Vüllers *et al.* [120]. Copyright 2018 American Chemical Society.)

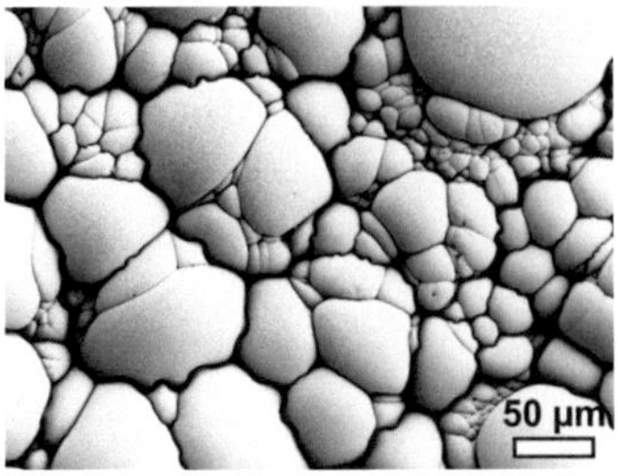

Figure 3.9: Negative PDMS imprint of nanofur.

Comparison of the Ni-shim (Fig. 3.8 b) to a negative PDMS imprint of thin nanofur (Fig. 3.9) shows that the electroformed structure is free of any defects.

After electroplating, the Ni-shim was glued to a steel plate and placed as top mold insert in a hot embossing machine above a 130 µm thick and 2 cm x 2 cm large FEP-foil resting on a PTFE sheet. After evacuating the embossing chamber, the shim was heated above the melting temperature of FEP ($T_m = 270\,°C$) and pressed into the FEP foil with an embossing force of $F_E = 10\,kN$. The system was then cooled down to room temperature and the structured FEP film was peeled of the shim without further treatment. Top view and cross-sectional SEM images of the embossed FEP MCA consisting of inverted hemispheres, previously observed on PC nanofur, are shown in Fig. 3.8 c), d). The Ni-shim was used for more than 70 consecutive replication cycles without showing any observable signs of wear. Additionally, as the replication procedure solely relies on hot embossing, it is not limited to FEP but can be transferred to a wide variety of polymers and even bulk metallic glasses in order to fulfill material restrictions imposed by special applications [90].

Furthermore, due to the low complexity of mold fabrication and replication process, the fabrication is feasible to be scaled up to a continous roll-to-roll fabrication technique. In order to demonstrate the defect-free large scale replication of random structures in fluorinated polymers on a roll-to-roll machine, FEP films were embossed with a sandblasted steel plate using a conventional embossing machine and with a sandblasted steel drum using a roll-to-roll embossing machine. Roll-to-roll embossing was carried out together with Andreas Striegel (IMT, Worgull Group) and SEM images of the respective embossed FEP films are compared in Fig. 3.10. Both replicated polymer films show a large scale waviness as well as small scale random protrusions of varying sizes. The identical topography of the two replication techniques acts as proof of concept for the scalability of FEP MCA fabrication.

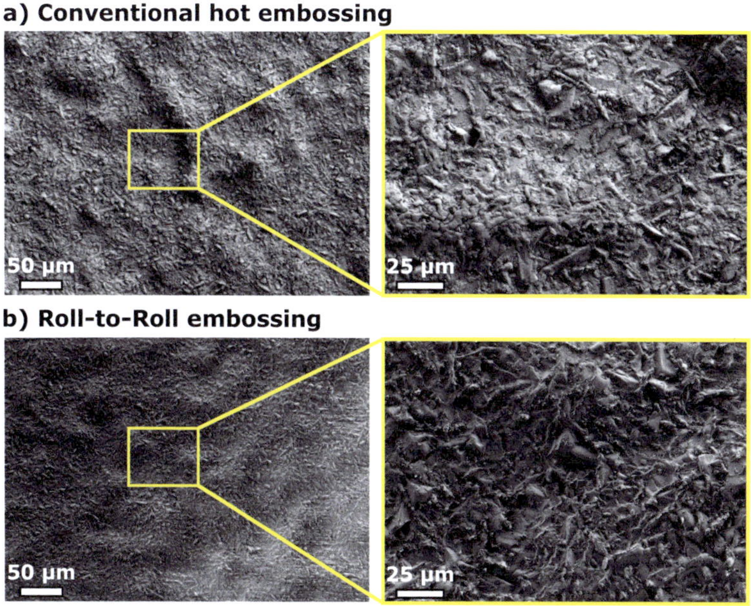

Figure 3.10: Upscaling of fabrication to a roll-to-roll process. a) SEM images of a sandblasted steel plate replicated into FEP with conventional hot embossing. b) SEM images of a sandblasted steel drum replicated into FEP with roll-to-roll embossing.

3.3 Fabrication of Surface Wrinkles and PTFE Pillars

To compare dynamic stability of the retained air layer of nanofur to other superhydrophobic surfaces and identify topographical features, that enhance the dynamic plastron stability, two additional superhydrophobic surfaces were fabricated: double scale teflon surface wrinkles (DS wrinkles) [123] and PTFE micropillars (PTFE pillars) [51]. SEM images of the surface topography of DS wrinkles and PTFE pillars are shown in Fig. 3.11.

DS wrinkles were fabricated through a self-assembly process following Scarratt *et al.* [123] in the lab of Chiara Neto at the School of Chemistry, University of Sydney. A polyethylene (PE) foil was sonicated in ethanol before being dried with high purity nitrogen gas. The foil was then placed in an oven at 110 °C and preshrunk around a 5 cm x 5 cm glass slide. Afterwards, the preshrunk foil was spincoated with 50 µl of 1.5 wt% Teflon AF solution at 3000 rpm for 60 s, before being removed from the glass slide and again placed in an oven at 110 °C until the film shrunk to approximately 1.5 cm x 1.5 cm. This treatment results in a threefold hierarchy of nanowrinkles on top of submicron wrinkles, which in turn fold into microwinkles (Fig. 3.11 a). The self-assembly process results in a random nature of the surface, which gives rise to a large variation in surface feature dimensions. The typical DS wrinkles sample consists of wrinkles with a width of 600 ± 230 nm and a heigth of 360 ± 210 nm and folds with 7.8 ± 2.3 µm width and 2.8 ± 1.1 µm height. The hierarchy of wrinkles additionally results in a cavity-like topography, as well as reentrant structures.

PTFE pillars were fabricated according to [51] by hot embossing (Section 2.3.1) a mold consisting of micropores into a PTFE foil with an embossing temperature of $T_E = 290\,°C$ and an embossing force of $F_E = 850$ kN. PTFE pillars are 21 µm tall, have a pitch of 20 µm and a tip diameter between 4 µm and 5 µm. Furthermore, unlike DS wrinkles, the PTFE pillars display no hierarchy (Fig. 3.11 b).

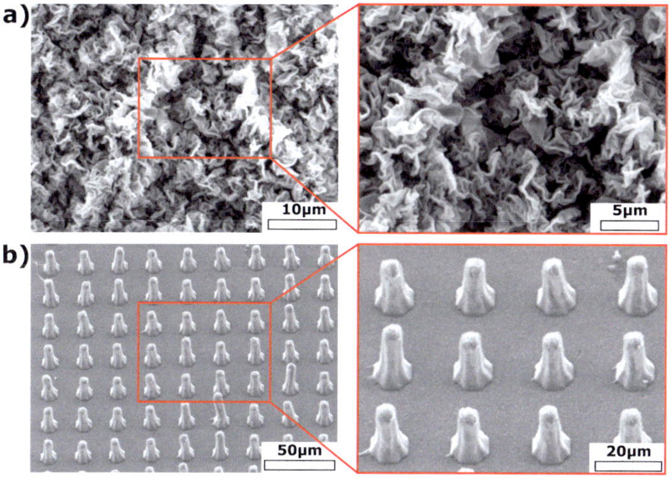

Figure 3.11: Surface topography of DS wrinkles and PTFE pillars. a) SEM images of DS wrinkles show the hierarchical wrinkle topography and microfolds forming cavities and reentrant structures. b) SEM images of highly ordered PTFE pillars display no hierarchical features.

3.4 Wetting Properties

3.4.1 Wetting Properties of Thin Nanofur

As described in Section 2.1, surfaces are classified as superhydrophobic if they exhibit a water contact angle $\theta_W > 150°$, which is essential to achieve self-cleaning properties. In this context, the wetting properties of thin nanofur were analyzed by measuring its water contact angle and water sliding angle as shown in Fig. 3.12 a), b). For water contact angle measurements a 3 µl DI water droplet was dispensed on the thin nanofur surface, while the water sliding angle was measured by dispensing a 10 µl DI water droplet, and subsequently tilting the surface until the droplet rolled off. Water droplets dispensed on thin nanofur keep an almost spherical shape and roll off at the smallest disturbances or vibrations of the surface. This superhydrophobic behavior is verified by a water contact angle of

$\theta_W = 166 \pm 6°$ and a water sliding angle $\theta_{WS} < 6°$. In contrast, unstructured PC is hydrophilic with a water contact angle of $72 \pm 3°$ and dispensed water droplets firmly stick to its surface. Because nanofur is not subjected to chemical treatment after the hot pulling process, the change from hydrophilicity (unstructured PC) to superhydrophobicity (thin nanofur) is solely due to the surface topography of thin nanofur with its high density of micro- and nanohairs. The hairs minimize the contact area between water and substrate, and, thus, trap air underneath the liquid. This entrapped air stops the liquid from penetrating into the topography, effectively forming a Cassie-Baxter wetting state [61, 124].
The influence of perforation and protective PDMS encapsulation on the wetting properties of thin nanofur was partially investigated in the master thesis of Meike Kronast (IMT) [125]. Manual perforation and laser drilling resulted in no change of wetting behavior. Plasma etching without protective cover significantly reduced the CA and increased droplet adhesion, while the PDMS encapsulation kept the wetting properties intact. These measurements indicate, that only plasma etching without a protective cover damages the nanofur surface topography, while PDMS encapsulation, laser and manual perforation do not alter the nanofur topography as indicated by the unchanged wetting properties.

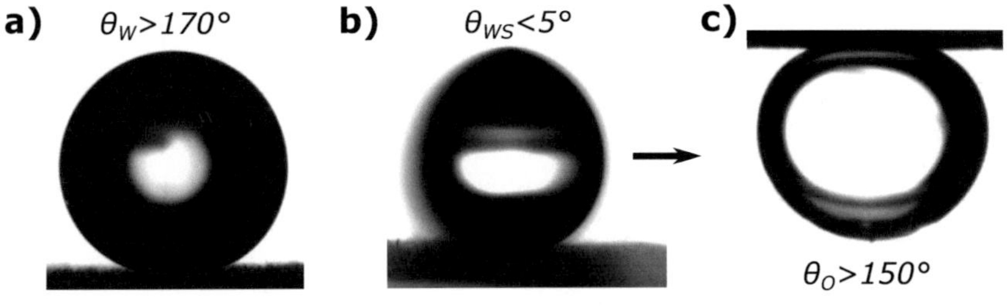

Figure 3.12: Wetting properties of thin nanofur. a) Photograph of a 3 µl water droplet on nanofur maintaining an almost ideal spherical shape with a water contact angle $\theta_W > 170°$ b) Photograph of a 10 µl water droplet sliding off the nanofur at a water sliding angle $\theta_{WS} < 6°$. c) Photograph of a 6 µl hydraulic oil droplet on a plasma-treated thin nanofur film submerged underwater with an oil contact angle $\theta_O > 150°$ and an oil sliding angle $\theta_{OS} < 2°$. (Figure reprinted with permission from Vüllers *et al.* [72]. Copyright 1999-2018 John Wiley & Sons, Inc.)

3.4.2 Reversing the Wetting Behavior

Even though superhydrophobicity is a highly desirable property for many applications, it can be necessary to modify the wetting behavior of extreme wetting films. Underwater superoleophobicity, which is beneficial for lossless oil droplet transport, lab-on-chip devices or oil-water separation [126, 127], requires high aspect ratio surface features with hydrophilic properties. Because high aspect ratio features are prevalent on most superhydrophobic surfaces, the combination of topography and wetting behavior required for underwater superoleophobicity can be achieved by reversing the wetting behavior of superhydrophobic surfaces. Plasma treatment is often used to precisely tailor the wetting behavior of polymers and metals [112, 128]. To reverse the wetting behavior of nanofur from superhydrophobic to hydrophilic, while keeping the surface topography intact, thin nanofur was exposed to argon plasma for 30 s at 30 W and 0.2 mbar. This treatment results in the formation of hydrophilic chemical groups on the surface [129], making the plasma treated nanofur (p-nanofur) underwater superoleophobic with an oil in water CA of $\theta_O > 150°$ (Fig. 3.12 c) and an oil sliding angle below $2°$. Because of the hydrophilicity of p-nanofur, water fills the space in between the micro- and nanohairs, thus, forming a separating water layer between the thin nanofur surface and the oil droplet. As a result, the oil droplet rests on a water-solid composite surface resembling an underwater Cassie-Baxter state, which results in excellent oil repellency [130]. Hence, the effect a water layer has on oil repellency is analogous to the effect the retained air layer of a superhydrophobic surface has on water adhesion.

However, the change in wetting behavior through plasma treatment is not permanent and the treated surface reverts back to its high CA through the so-called hydrophobic recovery [129, 131, 132]. Reorientation and mass diffusion of polar groups inside the material are responsible for hydrophobic recovery [129, 131].

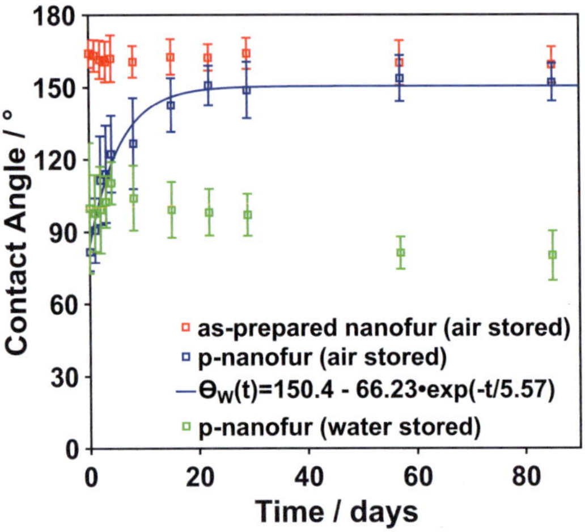

Figure 3.13: Hydrophobic recovery of p-nanofur. Average water contact angle as a function of storage time for as-prepared nanofur stored in ambient atmosphere and p-nanofur stored in ambient atmosphere and stored underwater. Error bars display the standard deviation from at least 7 measurements. Storage in water can delay or prevent hydrophobic recovery. (Figure adapted from Kavalenka *et al.* [78].)

However, recovery speed can be influenced by controlling the storage parameters of the plasma-treated surfaces, such as temperature, humidity or surrounding medium [129, 131, 132]. In order to study hydrophobic recovery and the effect storage conditions have on p-nanofur, the water contact angle of p-nanofur stored in ambient air and DI water was measured for 85 days. The measurements were compared to control samples of as-prepared thin nanofur stored in ambient air. The average measured CA as a function of storage time for water and air stored p-nanofur and as-prepared nanofur is shown in Fig. 3.13. P-nanofur stored in air recovers its high contact angle ($\theta_W = 154 \pm 7°$) after approximately three weeks of aging. The recovery is in good agreement with a limited growth model used by Bormashenko *et al.* [132] for unstructured surfaces:

$$\theta_W(t) = \widetilde{\theta}(1 - \exp(-t/\tau)) + \theta_O = \theta_{sat} - \widetilde{\theta}\exp(-t/\tau) \qquad (3.1)$$

where θ_O is the initial CA after plasma treatment, τ is the characteristic CA restoration time, $\widetilde{\theta}$ is a fitting parameter and $\theta_{sat} = \widetilde{\theta} + \theta_O$ is the saturation contact angle. Both $\widetilde{\theta}$ and τ need to be defined by fitting experimental data to Eq. (3.1). The average water contact angle for p-nanofur after plasma treatment is $\theta_W = 100 \pm 27°$ and in contrast to the air stored p-nanofur does not revert back to superhydrophobicity for water stored p-nanofur, but stays at $\theta_W = 97 \pm 4°$ for the first 30 days. After 30 days a significant drop in water contact angle ($\theta_W = 82 \pm 7°$) can be observed, which is most likely due to absorption of water into PC [132]. The results from Fig. 3.13 indicate, that hydrophobic recovery can be prevented by storing the plasma treated surfaces in water, thus, making it possible to preserve underwater superoleophobic properties of p-nanofur.

3.4.3 Wetting Properties of FEP MCA

Similar to nanofur, the wetting properties of FEP MCA were characterized by measuring water contact angle and water sliding angle. Figure 3.14 a) shows colored water droplets resting on unstructured FEP and FEP MCA. On unstructured FEP the droplet forms a hemisphere and firmly sticks to the substrate, while water dispensed on FEP MCA stays in an almost spherical shape and rolls off at the smallest disturbances, such as tilt or vibrations. Like most fluorinated polymers, FEP is highly hydrophobic with a water contact angle of $\theta_W = 110 \pm 1°$. However, water droplets dispensed on flat FEP firmly stick to the surface up to a tilt angle of approximately 41°, after which, they start to creep along the surface. In contrast, the increased roughness on FEP MCA enhances the polymers wetting properties [57, 62, 133], thus, resulting in a superhydrophobic surface with a water contact angle of $\theta_W = 158 \pm 6°$ and a low water sliding angle of $\theta_{WS} = 5 \pm 3°$ as shown in Fig. 3.14 b), c).

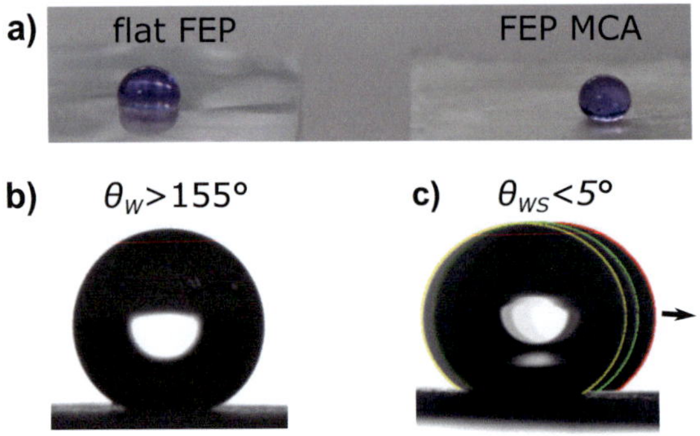

Figure 3.14: Wetting properties of FEP MCA. a) Colored water droplets resting on unstructured FEP and FEP MCA forming a hemisphere and a sphere respectively, thus, demonstrating hydrophobicity and superhydrophobicity of FEP and FEP MCA. b) Photograph of a 3 µl droplet resting on FEP MCA displaying a $\theta_W > 155°$. c) Time lapse series of a 12 µl water droplet rolling of FEP MCA at $\theta_{WS} < 5°$. (Figure reprinted with permission from Vüllers *et al.* [120]. Copyright 2018 American Chemical Society.)

The fabrication procedure and wetting properties of thin nanofur were published in "Bioinspired Superhydrophobic Highly Transmissive Films for Optical Applications", Felix Vüllers, Guillaume Gomard, Jan B. Preinfalk, Efthymios Klampaftis, Matthias Worgull, Bryce S. Richards, Hendrik Hölscher and Maryna N. Kavalenka, Small, 2016, 12, 6144-6152.

An extended investigation of hydrophobic recovery was published in "Adaptable bioinspired special wetting surface for multifunctional oil/water separation", Maryna N. Kavalenka, Felix Vüllers, Jana Kumberg, Claudia Zeiger, Vanessa Trouillet, Sebastian Stein, Tanzila T. Ava, Chunyan Li, Matthias Worgull and Hendrik Hölscher, Scientific Reports, 2017, 7, 39970-39979.

The fabrication procedure and wetting properties of FEP MCA were published in "Self-Cleaning Microcavity Array for Photovoltaic Modules", Felix Vüllers, Benjamin Fritz, Aiman Roslizar, Andreas Striegel, Markus Guttmann, Bryce S. Richards, Hendrik Hölscher, Guillaume Gomard, Efthymios Klampaftis and Maryna N. Kavalenka, ACS Applied Materials & Interfaces, 2018, 10, 2929-2936.

4 Stability of Retained Air Layer on Nanofur

Although nature has created many drag-reducing surfaces, engineering artificial drag-reducing surfaces is challenging to this date. Several different techniques, such as microbubble injection, sharkskin inspired riblets and polymer additives [14, 23–26], were developed to reduce frictional drag but demonstrated limited efficiency, cost-effectiveness or scalability [14, 27].
In nature, aquatic plants and insects utilize their superhydrophobic surfaces to retain an air film underwater and, thus, efficiently move in and on water [10, 11, 134, 135]. The lubricating air layer, also known as plastron, has been broadly investigated for its drag-reducing capabilities [12, 13, 81, 136]. Liquids flowing over a surface experience the no-slip boundary condition resulting in zero velocity at solid-liquid interfaces (Fig. 4.1 a). On the shear-free air layer on the other hand, the no-slip condition does not apply, resulting in a positive interfacial velocity of the flowing liquid and an up to 70% reduced frictional drag (Fig. 4.1 b) [26, 77, 111]. The major drawback of these drag-reducing surfaces is the instability of the retained air layer against external stimuli such as elevated pressure, pressure fluctuations or flow, as well as instability against prolonged immersion [12, 26]. These external stimuli can replace the air layer with water, and, thus, result in a loss of beneficial interfacial properties of the surface, such as superhydrophobicity and drag reduction. However, to effectively use air retaining surfaces in drag-reducing applications, the retained air layer of the superhydrophobic surfaces has to be stable against the aforementioned external stimuli.

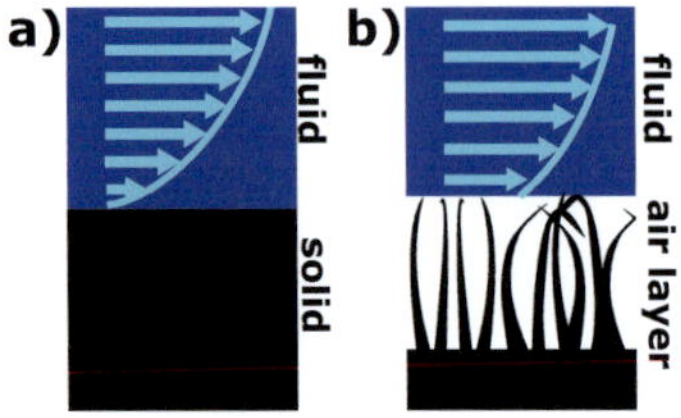

Figure 4.1: Drag reduction by lubricating air layer. Schematic illustration of the velocity profile on (a) an unstructured substrate and (b) an air retaining surface. The no-slip condition on the unstructured substrate results in zero velocity at the water-solid interface, while the lubricating air layer results in a positive velocity at the air-water interface.

This chapter focuses on overcoming these stability limitations of the retained air layer on thin nanofur in static conditions (Section 4.1) and identify plastron enhancing topographical features in dynamic conditions (Section 4.2). First the microscopic (Section 4.1.1) and macroscopic (Section 4.1.2) stability against hydrostatic pressure and the evolution of retained air layers on thin nanofur during wetting transitions is investigated. Then, a mechanism stabilizing the retained air layer against hydrostatic pressure is introduced and characterized in Section 4.1.3, before being extended to additionally stabilize the retained air layer against pressure fluctuations in Section 4.1.4. In Section 4.2.1, the dynamic stability of the retained air layer of three superhydrophobic surfaces, namely thin nanofur, DS wrinkles and PTFE pillars, is investigated and stability enhancing topographical features are identified. These topographical features are then correlated to the dynamic wetting properties and drag-reducing capabilities of the artificial superhydrophobic surfaces in Section 4.2.2. Finally, the stability of the three superhydrophobic surfaces against contamination is investigated in Section 4.2.3, followed by a demonstration of the directionality of dynamic wetting transitions in Section 4.2.4.

4.1 Retained Air Layer Stability Against Hydrostatic Pressure

Most natural and artificial superhydrophobic surfaces retain an air layer when submerged underwater, as indicated by a silver shimmer due to total internal reflection at the air-water interface (Fig. 2.10). As mentioned before, Solga *et al.* [68] identified the following criteria as essential for superhydrophobic surfaces with a stable long term underwater retained air layer: (i) hydrophobicity; (ii) micro- to millimeter long hairs; (iii) additional finer structures; (iv) micro- and nanocavities and (v) elastic structures. All these properties are prevalent on nanofur making it a promising material for stable air retention [81].

4.1.1 Air-Water Interface Evolution During Wetting Transition

As described in Section 2.1.3, wetting transitions can be induced through hydrostatic pressure and occur through two possible routes - sagging and sliding of the air-water interface [73–75]. In order to investigate the movement of the air-water interface during a pressure induced wetting transition, thin nanofur was placed in a pressure cell flooded with DI water and the hydrostatic pressure was increased in 100 mbar steps up to 1 bar (Fig. 4.2 a). At each pressure step, the air-water interface was then imaged using a confocal laser scanning microscope (CLSM) as described in Section 2.4.5. The air-water interface (light areas) spanning over as-prepared thin nanofur at hydrostatic pressures ranging from 0 mbar to 900 mbar is shown in Fig. 4.2 b). Black areas above the air-water interface resemble water resting on top of the nanofur, while the black area below the white regions corresponds to the retained air layer. Vertically connected black areas indicate wetted regions without a visible air-water interface. A custom Matlab script extracted the air-water interface position in each cross-sectional image, highlighted it with a red line and calculated the downward movement of the interface between pressure steps. With increasing water pressure, a distinct downward sliding and partial breakage of the air-water interface can be observed on as-prepared nanofur (Fig. 4.2 b).

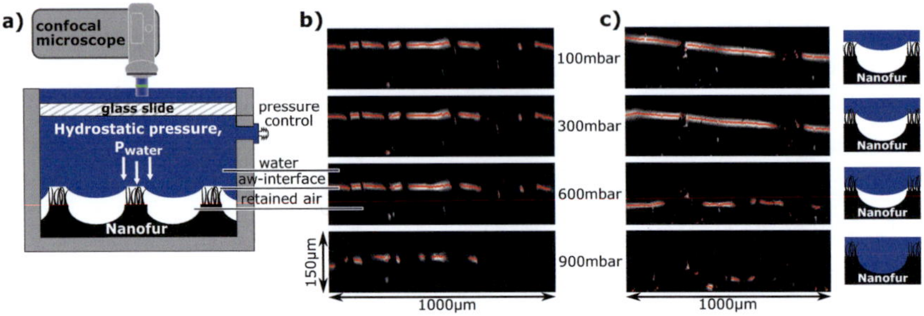

Figure 4.2: Wetting transition on superhydrophobic thin nanofur. a) Schematic illustration of measurement setup combining CLSM and a pressure cell. b), c) Series of cross-sectional vertical intensity profile images visualizes the air-water interface (white areas highlighted with red lines) suspended over (b) as-prepared and (c) silver sputtered thin nanofur. The air-water interface is interrupted by hairs from the surface topography and wetted areas appearing as breaks in the white and red lines. Increased hydrostatic pressure results in air-water interface sliding downward into the surface topography and partially breaking down. Right side shows a schematic illustration of the air-water interface at the respective hydrostatic pressures. (Figure adapted with permission from Vüllers *et al.* [137]. Copyright 1999-2018 John Wiley & Sons, Inc.)

After applying more than 900 mbar the air-water interface is completely broken down and the thin nanofur film is thoroughly wetted.

In order to differentiate between the effect of the topography consisting of a dense layer of hair and the effect of the inherent material wetting properties on the air-water interface behavior during wetting transitions, the air-water interface movement on thin nanofur films coated with high and low surface energy material was investigated and compared to the movement on as-prepared thin nanofur. To increase the surface energy of PC nanofur, a 150 nm thick silver layer was sputtered onto its surface, which reduces the water contact angle of unstructured PC from $\theta_{PC} = 80 \pm 4°$ to $\theta_{Silver} = 26 \pm 1°$. Coating surfaces with 1H,1H,2H,2H-perfluorodecyltriethoxysilane (FAS) on the other hand, reduces the surface energy significantly, as indicated by the increase in water contact angle of coated unstructured PC to $\theta_{FAS} = 101 \pm 11°$ [138]. Surprisingly, applied to thin nanofur, both, FAS and silver coating, slightly increase the water contact angle, as summarized in Table 4.1, rather than FAS increasing the contact angle and

Table 4.1: Static wetting properties and air-water interface behavior of as-prepared, silver-coated and FAS-coated nanofur under applied hydrostatic pressure. Low penetration depth and superior critical pressure of as-prepared nanofur with superhydrophobic surface topography and moderate surface energy compared to the same superhydrophobic nanofur topography with either low (FAS-coated) or high (silver-coated) surface energy indicates an increased plastron stability against hydrostatic pressure. (Table reprinted with permission from Vüllers *et al.* [137]. Copyright 1999-2018 John Wiley & Sons, Inc.)

Nanofur	CA unstructured	CA nanofur	Penetration depth	Interface breakdown
As-prepared	$80 \pm 4°$	$163 \pm 5°$	11 ± 5 µm	>900 mbar
Silver-coated	$26 \pm 1°$	$168 \pm 6°$	11 ± 5 µm	700 - 800 mbar
FAS-coated	$102 \pm 11°$	$173 \pm 4°$	92 ± 9 µm	500 - 600 mbar

silver decreasing it, as observed on unstructured PC. In the case of silver-coated nanofur, superhydrophobicity is most likely due to the unique surface topography and possible hierarchy of the complex silver layer sputtered on the rough surface [139]. Moreover, the minimal solid-water contact area of the dense layer of nano- and microhairs on nanofur possibly has a greater effect on the overall wetting behavior than the surface energy, resulting in a Cassie-Baxter wetting state, in spite of the high surface energy [139].

The downward movement and shape of the air-water interface was measured on all three nanofur types in respect to applied pressure. Surprisingly, sagging of the air-water interface was observed only on a few samples of as-prepared nanofur and on none of the coated nanofur samples. A low critical radius of the air-water interface as well as the low vertical resolution of the CLSM might be the reason for a lack of observable sagging. Table 4.1 compares the interface penetration depth, which describes the relative movement of the air-water interface from 0 mbar to 300 mbar applied pressure of the different types of nanofur. For each nanofur type the penetration depth was calculated from at least six vertical intensity profile series. On the first 300 mbar, as-prepared and silver-coated nanofur exhibit a similar penetration depth, whereas the penetration depth on FAS-coated nanofur is one order of magnitude higher than on the other two types (Tab. 4.1). The high penetration depth on FAS-coated nanofur is a result of the air-water interface resting on the apex of the highest microhairs in a metastable state. An increase in hydrostatic pressure results in an immediate downward movement of the air-water interface closer to the substrate and to the next layer of supporting hairs. Additionally, the increased movement is a result of easier air-water interface sliding due to the low surface energy of the material. In contrast, the hydrophilic hair tips on as-prepared and silver-coated nanofur can pin the air-water interface, thus, hindering its downward movement. For pressures exceeding 300 mbar however, the air-water interface on silver-coated nanofur slides downward faster than on as-prepared thin nanofur (Fig. 4.2 b), c)). The same trend prevalent in penetration depth can also be observed in the hydrostatic pressure required for completely breaking down the air-water interface. While the critical pressure required for a wetting transition on silver-coated nanofur is only slightly lower than on as-prepared nanofur, it is only half as much on FAS-coated nanofur as it is on as-prepared nanofur, indicating a reduced plastron stability.

These results indicate that a combination of superhydrophobic surface topography and presence of moderate surface energy material increases stability of the air-water interface against hydrostatic pressure compared to the same superhydrophobic topography with either high (silver-coated) or low (FAS-coated) surface energy. A possible explanation for this superiority is the moderate surface energy of as-prepared nanofur compared to the other two extreme surface energy cases of FAS-coated and silver-coated nanofur. On silver coated nanofur (high surface energy) the air-water interface is pinned at the apexes of the hairs, thus, stabilizing it. However, the high surface energy also results in an easier wettability of the nano- and mirohairs themselves, and, thus, a lower critical pressure than on as-prepared nanofur. On the other hand, on FAS-coated nanofur (low surface energy) a higher energy input is required to wet the nano- and microhairs themselves, but the missing pinning effect of the air-water interface at the tips of the hairs results in a lower stability. Therefore, the slightly hydrophilic properties of as-prepared nanofur result in a pinning of the air-water interface at the tips of the hairs, while simultaneously having a low enough surface energy to require a significant energy input for water to penetrate into the topography, resulting in a maximum critical pressure. This outcome is in agreement with observations on the leaves of water fern *Salvinia*, where a superhydrophobic hairy topography coupled with hydrophilic hair tips results in a retained air layer stable against high hydrostatic pressures [2].

4.1.2 Retained Air Layer Stability of Thin Nanofur

Section 4.1.1 investigated the microscopic behavior and small scale stability of the retained air layer on thin nanofur. However, nanofur is a highly heterogeneous surface and for many applications the macroscopic stability of the retained air layer is of importance. The next section investigates this macroscopic stability in more detail.

Hydrostatic Stability of Retained Air Layer on Nanofur

In order to investigate the robustness of the retained air layer on nanofur, $n = 10$ nanofur samples were placed in a pressure cell filled with DI water and the air-water interface was studied using total internal reflection as described in Section 2.4.4. After sealing the cell, the hydrostatic pressure was increased and consecutive *in-situ* images of the underwater retained air layer were taken at different applied pressures. The percentage of dry area was extracted from these images and normalized to the dry area without applied pressure. The normalized dry area as a function of applied pressure is shown in Fig. 4.3 a), with the inset displaying processed images of two typical measurements at different applied pressures.
The retained air layer decreases exponentially with increasing hydrostatic pressure with approximately 50% still intact at 0.5 bar, and approximately 10% at 1.4 bar. In comparison, water starts penetrating in between the eggbeater-shaped hairs of *Salvinia molesta* at ≈ 0.12 bar and completely wets the surface at 5.8 bar [4], while the lotus leaf can only sustain an air layer up to 0.135 bar [140]. Furthermore, to investigate directionality of pressure induced wetting transitions, the retained air layers on the upper and lower half of several thin nanofur films were processed and analyzed separately (Fig. 4.3 b). No difference in retained air layer stability

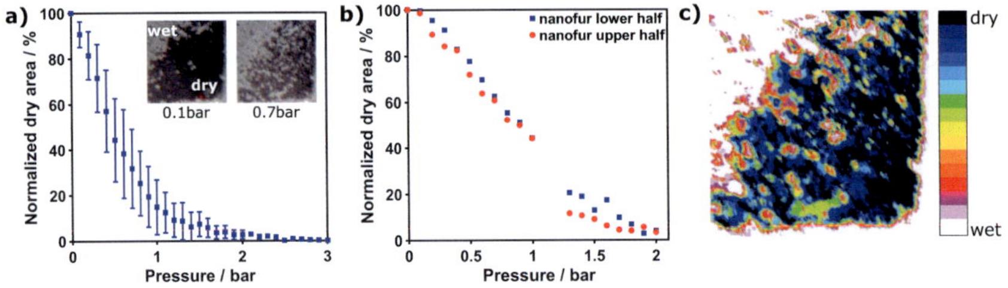

Figure 4.3: Retained air layer stability of nanofur. a) Normalized dry nanofur area averaged over $n = 10$ samples as a function of applied pressure. The inset shows processed images from two typical measurement at different applied pressures. The dry area appears bright, while the wetted area is dark. (Adapted with permission from Kavalenka *et al.* [81]. Copyright 2015 American Chemical Society.) b) Normalized dry area of one nanofur film divided into two sections as a function of applied pressure. The wetting behavior is identical for both halves, indicating the wetting transition to exhibit no global directionality. c) Overlaid images of intact air layer on superhydrophobic surfaces after pressurization. Wetting transitions spread outward from broken down areas, rather than form new broken down spots, indicating a local directionality.

between the lower and upper half can be observed, thus, indicating that hydrostatic pressure induced wetting transitions display no global directionality. Figure 4.3 c) shows overlapped images taken of the intact retained air layer of thin nanofur during exposure to hydrostatic pressure ranging from 0 bar to 1 bar. The merged images verify the absence of a global directionality (spreading from left to right or bottom to top) of the wetting transitions, as the retained air layer breaks down from all sides and independently of vertical position. Nonetheless, the images show that wetting transitions spread outwards from already broken down areas, rather than form new wetted areas. This indicates a local directionality and suggests, that the retained air layer stability is limited by the weakest area on the surface.

Calculated critical pressure

The experimental results are compared to the theoretically predicted critical pressure necessary to induce a wetting transition on thin nanofur. The critical pressure is calculated based on the pore model developed by Hemeda *et al.* (Section 2.1.3) [73], and thin nanofur is assumed to be comprised of cylindrical pores with uniform depth and diameter. The pore model is deemed fitting for nanofur because of the fabrication method utilizing a rough sandblasted steel plate to form microcavities surrounded by a layer of micro- and nanohairs (Fig. 3.2; Fig. 3.7). The applied model further assumes that the cavities are not interconnected and that each cavity has a diameter of 53 μm. The assumed cavity diameter is based on the blasting particle size used for sandblasting the mold insert. The depth of the cavities is determined from cross-sectional SEM images as 56 ± 19 μm. Inserting water surface tension $\gamma_W = 72.75\,\mathrm{mN\,m^{-1}}$ and ambient air pressure $p_{amb} = 1.03$ bar into Eq. (2.10) results in a theoretical critical pressure of $p_{cr,pore} = 0.43$ bar for thin nanofur. Furthermore, Fig. 4.4 shows the dependency of the critical pressure on pore diameter and pore depth, with the shaded area depicting the critical pressure for the measured cavity depth and its standard deviation. The calculated critical pressure is in good agreement with the experimentally derived value. Additionally, the experimentally observed exponential decrease in retained air layer can be explained by the variation in microcavity diameter on nanofur, resulting in different critical pressures for different sample areas.

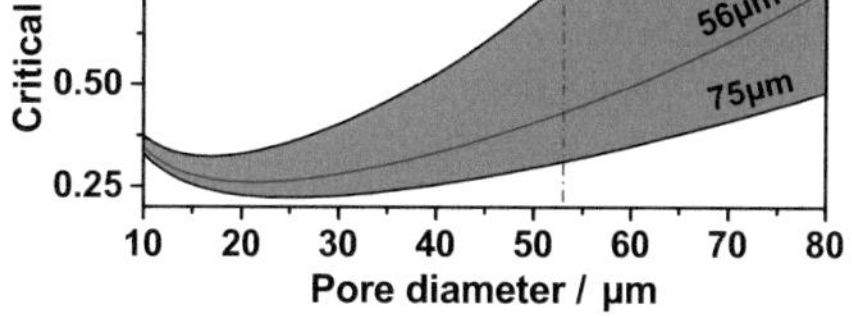

Figure 4.4: Theoretical critical pressure of thin nanofur. Calculated critical pressure as a function of pore diameter and depth. The shaded area depicts the critical pressure calculated for the measured pore depth and its standard deviation. The assumed pore diameter is highlighted. (Reprinted with permission from Kavalenka *et al.* [81]. Copyright 2015 American Chemical Society.)

4.1.3 Highly Pressure-Stable Air Layer of Perforated Nanofur

As described in Section 2.1.3, pressure induced wetting transitions are due to an imbalance of forces acting on the air-water interface. Therefore, to overcome the limitation in pressure stability of the retained air layer, an elevated hydrostatic pressure needs to be compensated by an increased air pressure. In nature, aquatic insects such as the fishing spider *Anclyometes bogotensis* and the water strider *Gerris remigis* can retain a stable air layer under pressures of up to 4 bar (corresponding to 40 m water depth) [134, 141]. This pressure stability is achieved through a combination of high hair density (6.6 million hairs per mm^{-2}), balancing water pressures with the retained air layer and increased oxygen diffusion from water into the retained air layer [134, 141]. To gain access to the underwater retained air layer on thin nanofur films, perforated nanofur was fabricated as described in Section 3.1.3. The access made it possible to adapt the natural mechanism of compensating hydrostatic pressures acting on the underwater retained air layer through increased retained air pressure. To this effect, the perforated nanofur was attached as top seal to a pressure cell, which was in turn connected to a compressor, and submerged in a second pressure cell partially flooded with DI water as shown in Fig. 4.5 a), b), thus, forming a double-cell system. The individual connection of both cells to the compressor allows precise control over the hydrostatic pressure (p_{water}) acting on the air-water interface and the pressure in the retained air layer (p_{air}). As a result, the hydrostatic pressure acting on the air-water interface can be compensated by the pressure supported retained air layer, thus, increasing plastron stability. First the stability of the air-water interface subjected to equally high pressures from both sides was tested. To do this, the evolution of the retained air layer was investigated using total internal reflection as described in Section 2.4.4. After submersion of perforated nanofur in the double-cell setup, the valves connecting the retained air layer and the outer pressure cell to the compressor were opened, thus, simultaneously increasing both hydrostatic and retained air pressure ($p_{water} = p_{air}$). During the experiment the pressures were monitored using a pressure gauge. Consecutive photographs of the air-water interface were taken at

500 mbar intervals and the dry area was extracted and normalized to the dry area with no applied pressure, thus, making a comparison between pressure-supported perforated nanofur (pp-nanofur) and as-prepared nanofur possible. The averaged normalized dry area of pp-nanofur ($n = 14$) as a function of applied pressure is compared to that of as-prepared thin nanofur in Fig. 4.5 c). At 1.5 bar the normalized dry area of pp-nanofur drops to 88% and then stabilizes at 86% with pressure increasing to 4 bar (corresponding to a water depth of 40 m). In contrast, the normalized dry area of as-prepared nanofur decreases exponentially with only 7.2% still intact at 1.5 bar and the surface completely wetted at 2.2 bar.

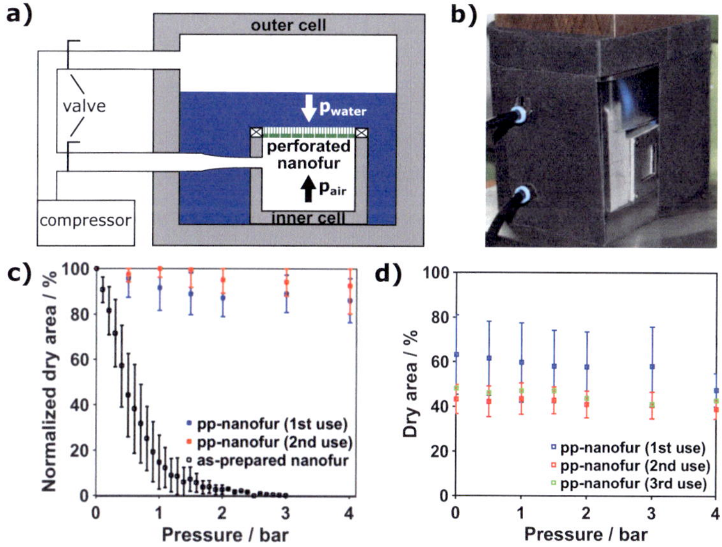

Figure 4.5: Pressure stability of pressure-supported perforated nanofur (pp-nanofur). Schematic (a) and photograph (b) of the double-cell setup used to support and analyze the retained air layer on perforated nanofur. Attaching nanofur to the inner cell and connecting both the inner and outer cell to a compressor allows precise control over water and retained air pressure. c) Averaged normalized dry area as a function of applied pressure for as-prepared, pp-nanofur and reused pp-nanofur. Error bars correspond to the standard deviation. d) Overall dry area as a function of applied pressure for pp-nanofur used for the first, second and third time. (a)-c) reprinted with permission from Vüllers *et al.* [137]. Copyright 1999-2018 John Wiley & Sons, Inc.)

Furthermore, four pp-nanofur samples were subjected to 7.2 bar pressure and still retained an average normalized dry area of 85% (not shown here). Therefore, stability of pp-nanofur is increased by several orders of magnitude compared to as-prepared nanofur. Due to mechanical limitations of the pressure cells and pressure limitations of the installed compressor no higher hydrostatic pressures could be tested. However, as no significant degradation could be observed by increasing pressure from 1.5 bar to 7.2 bar, the retained air layer on pp-nanofur can reasonably be assumed to be stable at even higher pressure. The stability enhancement can be explained by regarding the balance of forces (eq. 2.5) acting on the underwater retained air layer. By compensating the hydrostatic pressure with the supported retained air pressure ($p_{water} = p_{air}$) the balance of forces simplifies to $p_{air} + p_{cap} - p_{water} - p_{amb} = p_{cap} - p_{amb} = 0$. Thus, the pressure support results in a situation equal to no applied hydrostatic pressure, which is highly stable. Moreover, Fig. 4.5 c) shows, that even during repetitive exposure to high hydrostatic pressure, more than 85% of normalized dry area stays intact, thus, making pp-nanofur reusable without significant loss of interfacial properties. The slight increase in average normalized dry area of pp-nanofur used for the second time compared to pp-nanofur used for the first time is due to metastable areas of pp-nanofur irreversibly breaking down during the first use (Fig. 4.5 d). As a result, the total dry area does not recuperate from the first to the second use, but does recuperate between the second and third use, indicating that areas intact after one pressurization can recover from a subsequent wetting transition.

Besides being stable under elevated hydrostatic pressure, the retained air layer on superhydrophobic surfaces utilized in drag-reducing application is required to be stable for long periods of immersion. However, as shown by previous studies, increased hydrostatic pressure results in a significant lifetime reduction of the retained air layer [27, 73, 142]. The effect of elevated hydrostatic pressure on plastron lifetime was investigated by exposing pp-nanofur and as-prepared nanofur to hydrostatic pressure for up to 4 hours. The normalized dry area as a function of immersion time for both films under different applied pressures is shown in Fig. 4.6. On pressurized as-prepared nanofur the dry area decreases exponentially with time. After two hours at 0.1 bar and 0.3 bar the dry area decreases to 55% and 35% respectively, demonstrating an increased decay of dry area through increased pressure. In contrast, the retained air layer on pp-nanofur pressurized up to 3.5 bar is stable for more than four hours, after which the experiment was stopped. Increasing the pressure to 4 bar, surprisingly results in a complete wetting transition of the nanofur surface after two hours. The plastron lifetime of pp-nanofur is in the same order of magnitude as that of surfaces fabricated by Samaha *et al.* [143], who have recorded an air layer stable under 7.5 bar for up to 3.5 h on a fibrous polystyrene coating electrospun onto an aluminum plate. In comparison, most other artificial superhydrophobic surfaces can sustain a stable retained air layer at up to 130 mbar for no more than 4 hours, with longevity decreasing exponentially with increasing pressure [144–146]. Natural surfaces such as the lotus leaf break down after just 1 hour at 55 mbar [147].

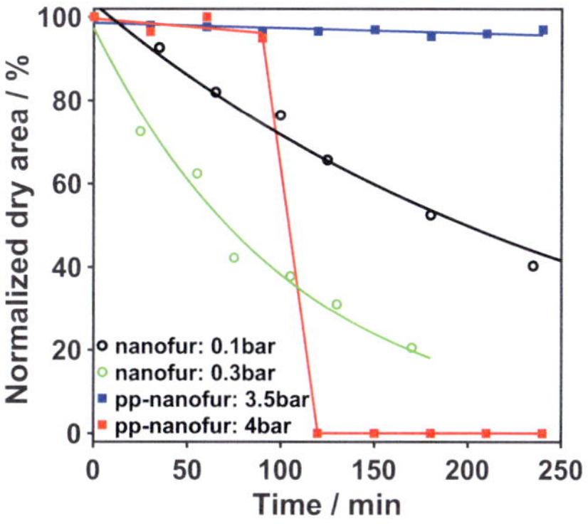

Figure 4.6: Plastron lifetime on pp-nanofur. Normalized dry area as a function of immersion time for as-prepared nanofur and pp-nanofur. Lines represent fits to the respective data. Lifetime of the retained air layer is significantly increased by pressure support. (Figure reprinted with permission from Vüllers *et al.* [137]. Copyright 1999-2018 John Wiley & Sons, Inc.)

Influence of perforation parameters on air layer stability

To study the influence pore location has on the stability of the retained air layer, thin nanofur was divided into three areas (named A, B, C) with varying pore positions as shown in the inset of Figure 4.7 a). The left half of the thin nanofur sample (area A) was perforated 20 times in the upper half of the area, while the top right quadrant (area B) was perforated 20 times over the whole area and the lower right quadrant (area C) was not perforated at all. To prevent cross-stabilization of pressure between the three areas, the superhydrophobic structure on their respective borders was mechanically removed (by abrasion with a pair of tweezers), thus, isolating the three areas and their respective retained air layers from each other. This isolation enables separate stability measurements of the retained air layer of each individual area. Figure 4.7 a) shows the divided thin nanofur film after exposure to 4 bar hydrostatic pressure. The bright color of both areas A and B shows, that the perforated areas stayed completely dry. The non-perforated area C on the other hand, is completely wetted after being subjected to hydrostatic pressure, as indicated by the darker color. Moreover, because both areas A and B stayed completely dry, despite area A only being perforated in the upper half, it can be concluded that the retained air layer stability is independent of pore positioning. In particular, dryness of the lower half of area A indicates, that the nano- and microstructure of nanofur is interconnected, resulting in a continuous retained air layer spanning over the complete left half of the sample. Furthermore, because the retained air layer is interconnected, pressure support is not localized but distributed over the complete retained air film. As a result, there is almost no limit to the distance over which a single pore can stabilize a retained air film. The only stabilizing barrier for pores are areas completely isolated by damaged superhydrophobic structure, and, thus, cut off from the pressure support, as can be seen in the varying stability of areas B and C.

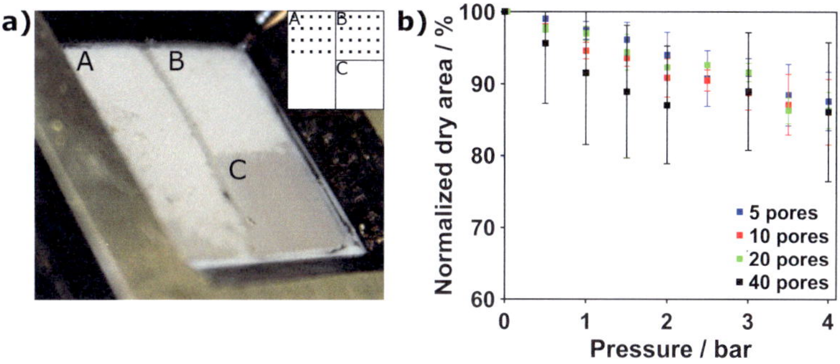

Figure 4.7: Influence of perforation parameters on retained air layer stability. a) Photograph of thin nanofur film divided into 3 areas (A-C) after exposure to 4 bar hydrostatic pressure. The three areas were separated by mechanically removing the superhydrophobic structure on their respective borders. The inset schematically shows pore location on the nanofur sample: A - only top half perforated; B - completely perforated; C - not perforated. Perforated areas (A,B) are dry (bright), while non-perforated area (C) is completely wetted (dark), indicating retained air layer stability to be independent of pore positioning. b) Normalized dry area as a function of applied hydrostatic pressure and number of pores. Stability of the retained air layer is independent of number of pores. (Figure reprinted with permission from Vüllers *et al.* [137]. Copyright 1999-2018 John Wiley & Sons, Inc.)

Furthermore, the influence of pore density on the stability of the retained air layer was investigated. The top half of several nanofur samples was perforated with 5, 10, 20 or 40 pores, before measuring the normalized dry area as a function of applied pressure up to 4 bar. Figure 4.7 b) shows that up to 4 bar, more than 86% of the retained air layer stays intact for all pore densities, indicating plastron stability to be independent of number of supporting pores. Therefore, it seems beneficial to reduce the number of pores to a minimum, as it reduces handling and processing times of the sample during perforation, and, thus, the possibility of damaging the superhydrophobic structure. However, as pores are only able to support dry areas connected to them (Fig. 4.7 a), too few pores might result in retained air layers isolated from a supporting pore, and, thus, not being stabilized. Hence, a compromise between reducing potential handling damages and ensuring a complete support of the retained air layer has to be found for practical applications. Minimizing damage to the superhydrophobic structure can also be achieved by reducing pore diameter and direct contact to the superhydrophobic surface during perforation, as is the case for laser drilling pores. To investigate the suitability of laser drilling for pressure support, the plastron stability of laser perforated and manually perforated thin nanofur was compared, and no difference in plastron stability was observed between the two perforation techniques.

4.1.4 Retained Air Layer Stability Against Water Pressure Fluctuations

Materials used in drag-reducing applications do not only experience static pressure but are often subjected to fluctuations of water pressure, which can result in water penetrating into the surface topography and a subsequent breakdown of the air-water interface. The previous section has shown that as long as the water pressure is balanced by an equivalent retained air pressure, the air-water interface is highly stable. By increasing the supporting air pressure above the hydrostatic pressure, sudden surges/fluctuations in hydrostatic pressure can be compensated, and the total hydrostatic pressure remains always below or at the same level as the retained air pressure. The effect such a supporting air pressure exceeding the hydrostatic pressure has on the stability of the air-water interface was studied on pp-nanofur. To this end, pp-nanofur was submerged in the double-cell setup and the hydrostatic (p_{water}) and retained air pressure (p_{air}) were simultaneously increased to a base pressure (p_{base}) ranging from 0 bar to 4 bar. After reaching p_{base}, the valve connecting the hydrostatic pressure to the compressor was closed, fixing $p_{water} = p_{base}$. Next, only p_{air} was further increased in 0.2 bar steps (Δp) up to an additional 3 bar. At each pressure step $p_{base} + \Delta p$ the normalized dry area of the retained air layer was extracted. To do this, first the retained air layer was depressurized back to p_{base}, before taking a photograph of the air-water interface. Depressurization of the retained air layer is necessary, as otherwise bubble formation would prevent an accurate depiction of the retained air layer. The average normalized dry area of three pp-nanofur films as a function of pressure difference $\Delta p = p_{air} - p_{base}$ is shown in Fig. 4.8 a). For low base pressure

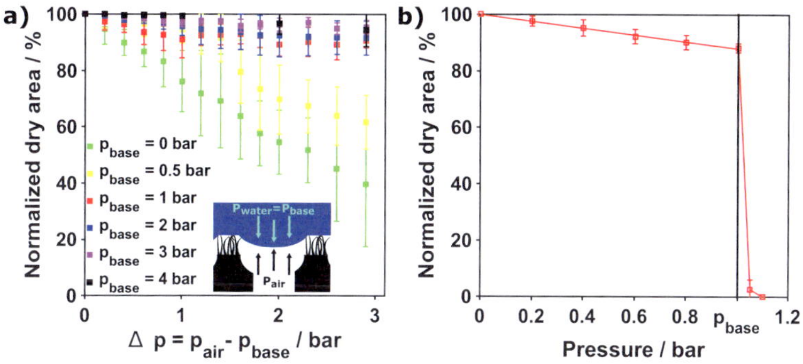

Figure 4.8: Stability of retained air layer against pressure fluctuations. Normalized dry area of pp-nanofur as a function of pressure difference between retained air pressure (p_{air}) and hydrostatic pressure (p_{water}). Air and water pressure were simultaneously increased to a base pressure (p_{base}), before only (a)p_{air}/ (b)p_{water} was additionally increased past p_{base}. a) With increasing p_{base} the air layer becomes more stable against increased pressure differences, creating a buffer pressure against pressure fluctuations. b) Increasing p_{water} past p_{base} results in an immediate wetting transition. (a) reprinted with permission from Vüllers *et al.* [137]. Copyright 1999-2018 John Wiley & Sons, Inc.)

($0\,\text{bar} \leq p_{base} \leq 0.5\,\text{bar}$) an increase in p_{air} results in a linear decrease in normalized dry area, while for $p_{base} \geq 1$ bar, the normalized dry area stabilizes at 89% up to an additional 3 bar buffer pressure. Additionally, overall plastron stability is shown to increase with increasing p_{base}. One possible explanation for reduced stability at low p_{base} is the formation of gas bubbles at the nanofur surface for p_{air} exceeding p_{base} by more than 50 mbar, which is of the same order of magnitude as the capillary pressure calculated for the thin nanofur topography. The formation and detachment of gas bubbles at the nanofur surface can lead to high local pressure fluctuations, and, thus, a wetting transition. These pressure fluctuations have a higher impact on plastron stability for low base pressure, due to the significantly higher relative pressure difference between pressure fluctuations and base pressure than for high p_{base}.

Additionally, by using the same measurement technique but increasing hydrostatic pressure above retained air pressure ($p_{base} = p_{air}$) the impact of hydrostatic pressure exceeding the supporting air pressure was investigated. As soon as $p_{water} > p_{air}$, pp-nanofur undergoes a complete wetting transition. This is in contrast to the expected exponential decay as observed on as-prepared nanofur (corresponding to $p_{base} = p_{air} = 0$ bar) shifted by p_{base}. However, the reason for the instantaneous wetting transition is the near infinite air volume added to the retained air layer by the connected inner pressure cell. Before perforation, increased water pressure compressed the retained air layer, resulting in an increased air pressure, which partially compensates the elevated hydrostatic pressure. With the added infinite air volume of the inner pressure cell and compressor, the downward shift of the air-water interface does not significantly change the volume of the retained air layer, and, therefore also does not change the retained air pressure. Thus, the retained air pressure is effectively eliminated from the balance of forces (Eq. (2.5)), leaving the capillary pressure as sole compensating force against elevated hydrostatic pressures, as described in Section 2.1.3. Therefore, once the hydrostatic pressure rises above the capillary pressure a complete wetting transition takes place as observed in Fig. 4.8.

4.2 Dynamic Plastron Stability of Superhydrophobic Surfaces

4.2.1 Retained Air Layer Stability of Moving Superhydrophobic Surfaces

In addition to high hydrostatic pressure, surfaces used for drag-reducing applications are exposed to movement through liquids or flow of liquids over the surface. Immersion and movement of a superhydrophobic surface into and through a liquid results in shear forces acting on the retained air layer, which, similar to the case of a liquid flowing over the surface, can result in a wetting transition, and, thus, a loss of interfacial properties. Several studies have investigated the effect liquid flow has on the longevity of the retained air layer [111, 148–151], and indirectly measured the dynamic plastron stability by detecting failure of drag-reducing properties [26, 136, 152]. Nonetheless, to my knowledge, no study to date has investigated the effect that immersion of a superhydrophobic surface into a liquid

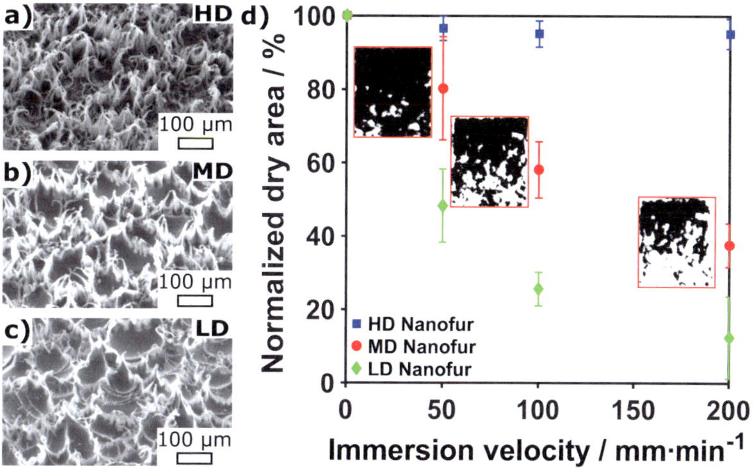

Figure 4.9: Dynamic plastron stability on nanofur. SEM images of a) high density (HD), b) medium density (MD) and c) low density (LD) nanofur. d) Normalized dry area of HD, MD and LD nanofur as a function of immersion velocity in DI water. The inset shows typical processed and thresholded images of MD nanofur at the respective immersion velocities, visualizing the dry (dark) and wetted (white) areas. A decrease in hair density and increase in cavity diameter reduces dynamic plastron stability.

has on the stability of the retained air layer. To investigate the dynamic stability of the retained air layer on nanofur, samples were attached to a dip-coater and submerged in DI water at velocities ranging from 20 mm min^{-1} to 200 mm min^{-1} (Fig. 2.8 a), b)). The dry area after each subsequent immersion was measured as described in Section 2.4.4 and normalized to the dry area prior to the first immersion. Three different classes of nanofur, varying in hair density and microcavity diameter were fabricated as described in Section 3.1.2 and tested: high density (HD), medium density (MD) and low density (LD) nanofur (Fig. 4.9 a) - c)).

The normalized averaged dry area of HD, MD and LD nanofur as a function of immersion velocity is shown in Fig. 4.9 d). During the first immersion at 50 mm min^{-1} the normalized dry area on HD nanofur reduces to 92% and then stabilizes at this value up to an immersion velocity of 200 mm min^{-1}. In contrast, the normalized dry area on MD and LD nanofur decreases exponentially with increasing dip speed and only 37% and 12% of the respective dry area is still intact after an immersion at 200 mm min^{-1}. Differences in topography are the reason for the accelerated decay of dry area on MD and LD nanofur compared to HD nanofur. The significantly higher hair density of HD nanofur compared to the other nanofur classes results in a lower pitch between nano- and microhairs supporting the air-water interface. Thus, the dynamic stability of the retained air layer against water penetration is increased. A similar behavior is observed for the retained air layer stability against hydrostatic pressure of superhydrophobic surfaces with high pitch [73]. The increased pitch enables the air-water interface to sag further into the cavities, and, thus, easier wet the surface. Additionally, the tip diameter of micro- and nanohairs on MD and LD nanofur is higher compared to HD nanofur, which results in a higher solid-water contact area, and, thus, a further reduction in dynamic plastron stability.

Next, double scale teflon surface wrinkles (DS wrinkles) and PTFE pillars were fabricated as described in Section 3.3 and the dynamic plastron stability on both surfaces was characterized. This enables a comparison of the dynamic plastron stability on nanofur to superhydrophobic surfaces ranging over multiple length scales, surface energies and structure regularity, and, thus, identify topographical features necessary for a dynamically stable retained air layer. Figure 4.10 a), b) shows SEM images of fabricated DS wrinkles and PTFE pillars. The DS wrinkles are comprised of a threefold randomly oriented wrinkle-hierarchy consisting of nanowrinkles on top of submicron wrinkles, which in turn fold into microwrinkles, and, thus, form cavity-like structures [123]. In contrast, PTFE pillars display no hierarchy, have a pitch of 20 µm, a diameter of 4 µm - 5 µm and a height of 21 µm [51]. The averaged normalized dry area of DS wrinkles, PTFE pillars and HD nanofur as a function of immersion velocity is shown in Fig. 4.10 c). On DS wrinkles the retained air layer is not affected by immersion velocity and stays 100% intact up to 200 mm min^{-1}, whereas, the retained air layer on the PTFE pillar surface behaves similarly to MD nanofur. It decays rapidly with increasing immersion velocity and only 37% is still intact after an immersion at 200 mm min^{-1}. As already observed on MD and LD nanofur, one

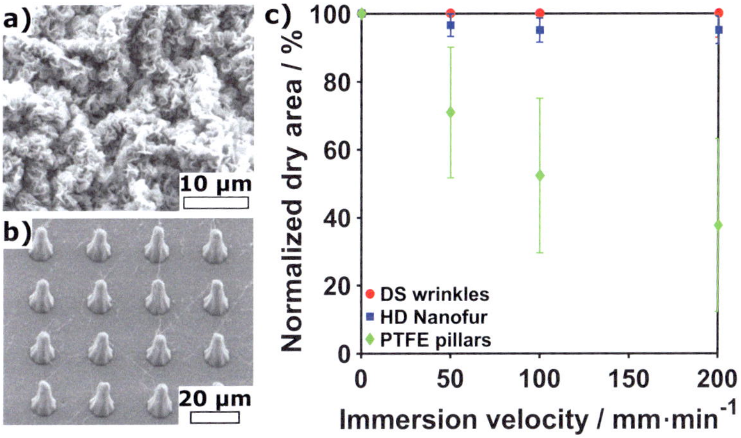

Figure 4.10: Dynamic plastron stability on DS wrinkles and PTFE pillars. SEM images of (a) DS wrinkles and (b) PTFE pillars. c) Averaged normalized dry area of DS wrinkles, HD nanofur and PTFE pillars as a function of immersion velocity in DI water.

of the reasons for the increased plastron decay on the PTFE pillar surface is the high pitch of the micropillars supporting the air-water interface compared to the densely packed features on HD nanofur and DS wrinkles. Additionally, the solid-water contact area is significantly higher for the PTFE pillar surface than for the nano- and microhairs and nanowrinkles on nanofur and DS wrinkles, which further reduces the dynamic plastron stability. This correlation between larger solid-water contact area and reduced plastron stability is in agreement with previous studies, in which nanoscale features were shown to exhibit an increased retained air layer stability against hydrostatic pressure and flow compared to larger microstructures [77, 111, 136, 153].

Another topographical difference between PTFE pillars, DS wrinkles and HD nanofur is the interconnected and open air layer on PTFE pillars compared to the cavities and cavity-like surface features prevalent on nanofur and DS wrinkles. In contrast to an "open" air layer, the cavities partially surround and enclose the retained air layer, thus, effectively forming a barrier against flowing water shearing of the air layer, and, as a result making the retained air layer less susceptible to movement. Moreover, the higher surface energy of PC nanofur in comparison to PTFE, possibly results in pinning of the air-water interface at the tips of the nano- and microhairs, further increasing stability of the retained air layer compared to the hydrophobic PTFE pillars. On the highly hydrophobic DS wrinkles the lack of pinning apexes is compensated by the nanometer scale feature size, threefold hierarchy and prevalent reentrant structures. The stability enhancing properties of hydrophilic apexes are in agreement with the results from Section 4.1.1 and observations on the water fern *Salvinia molesta*, on which a combination of wax coated superhydrophobic hairs and hydrophilic hair tips was shown to significantly increase the stability of the retained air layer [2]. Finally, DS wrinkles and HD nanofur are comprised of a hierarchical topography, which was shown to enhance superhydrophobicity and stability of the retained air layer against hydrostatic pressure, and, therefore, most likely further enhances the dynamic stability of the retained air layer on these surfaces, while the lack of hierarchy on PTFE pillars diminishes its dynamic plastron stability.

In summary, the dynamic stability measurements identify the following set of topographical features as necessary to achieve a high dynamic plastron stability: (i) hierarchical topography; (ii) high density surface features; (iii) minimal water-solid-contact area; (iv) nanoscale feature size; and (v) cavities or cavity-like structures.

4.2.2 Contact Line Motion and Dynamic Wetting Behavior

As mentioned before, application of superhydrophobic surfaces, especially in the case of drag reduction, usually exposes them to flow or movement through liquids, rather than static conditions. Even though dynamic behavior is essential for applications and studies have shown great variations between static (sessile drop technique) and dynamic (contact line motion) wetting behavior [108, 154–157], only static measurements are typically employed to characterize the wetting behavior of superhydrophobic surfaces. Therefore, application oriented characterization of wetting properties needs to include dynamic wetting behavior. The dynamic wetting behavior of HD nanofur, DS wrinkles and PTFE pillars was measured using a newly developed contact line motion technique described in Section 2.4.3 with an immersion velocity of 50 mm min^{-1}. Table 4.2 compares the static (sessile drop technique) and dynamic (contact line motion) wetting behavior of air retaining areas on all three used superhydrophobic surfaces. For the dynamic CA only surface areas with an intact air layer were included. Measurements with the sessile drop technique show almost no difference between DS wrinkles and PTFE pillars, with advancing and receding CAs of $172 \pm 2°$ and $170 \pm 2°$ for DS wrinkles [123] and $174 \pm 5°$ and $170 \pm 7°$ for PTFE pillars. HD nanofur has a slightly lower advancing and receding CA of $164 \pm 3°$ and $159 \pm 5°$. However, from the sessile drop technique no conclusions about the dynamic plastron stability can be drawn. Contact line motion on the other hand, shows DS wrinkles and HD nanofur to have an intact air layer with homogenous superhydrophobic wetting behavior for the complete structured area, whereas the PTFE pillar surface displays wetting transitions on part of the structured area, indicating a reduced dynamic plastron stability. Therefore, in contrast to the sessile drop technique, the *in-situ* contact line motion measurement can reveal differences in the dynamic plastron stability of the various surfaces.

HD nanofur displays the smallest difference ($\approx 5°$) between static and dynamic wetting behavior for both advancing and receding CA. Moreover, in contrast to the other surfaces, the dynamic advancing and receding CA of HD nanofur is above 150°, resulting in a minimal CAH. In contrast, the difference between static and dynamic wetting behavior for DS wrinkles and PTFE pillars is 20° for the advancing CA and 30° for the receding CA. This not only results in a drop in CA below 150° but also a tripling in CAH compared to the static wetting properties. This large difference in dynamic CAH between nanofur, DS wrinkles and PTFE pillars is surprising, as the static measurements show minimal CAH for all three surfaces. However, previous studies have shown that such a decrease in dynamic CA and increase in CAH indicates a reduced slip length [156, 158, 159]. Because slip length increases with (i) increasing plastron volume and

Table 4.2: Static and dynamic wetting properties of superhydrophobic surfaces measured using sessile drop method and contact line motion respectively. [a]Static wetting properties of DS wrinkles are taken from Scarratt *et al.* [123].

Surface	Sessile Drop		Contact Line Motion		Roll-off angle
	Advancing CA	Receding CA	Advancing CA	Receding CA	
PTFE pillars	$174 \pm 5°$	$170 \pm 7°$	$150 \pm 4°$	$142 \pm 4°$	$< 4°$
HD nanofur	$164 \pm 3°$	$159 \pm 5°$	$159 \pm 3°$	$155 \pm 6°$	$< 4°$
DS wrinkles	$172 \pm 2°$[a]	$170 \pm 2°$[a]	$151 \pm 4°$	$141 \pm 7°$	$< 4°$

(ii) decreasing solid-water contact area [111, 136], the increased CAH is a result of differences in the surface topography and can be explained by the small retained air volume on DS wrinkles and the relatively high solid-water-contact area of PTFE pillars compared to HD nanofur. Hence, HD nanofur is the most promising drag-reducing topography of the three investigated surfaces.

The combination of high dynamic stability and increased CAH of DS wrinkles compared to HD nanofur show that topographical features increasing the dynamic stability of the retained air layer can also decrease the drag-reducing capabilities of a superhydrophobic surface. Therefore, applications need to find a compromise between high dynamic plastron stability and high slip length, specifically tailored to the requirements of the application. This trade off between plastron stability and slip length is in agreement with previous studies, showing nanometer-scale roughness to enhance the stability of a retained air layer against elevated hydrostatic pressure, but simultaneously reducing the surfaces slip length compared to microstructured surfaces [111, 136, 153].

4.2.3 Contamination Resistance of Superhydrophobic Surfaces

Real world application of air retaining surfaces often sees them exposed to liquids other than DI water. Immersion in such liquids can lead to bio-fouling or other contamination of the superhydrophobic surface [6, 160]. Biofouling is typically characterized by prolonged immersion of superhydrophobic surfaces in sea water or bacterial solutions, including exposure to flow [160, 161].

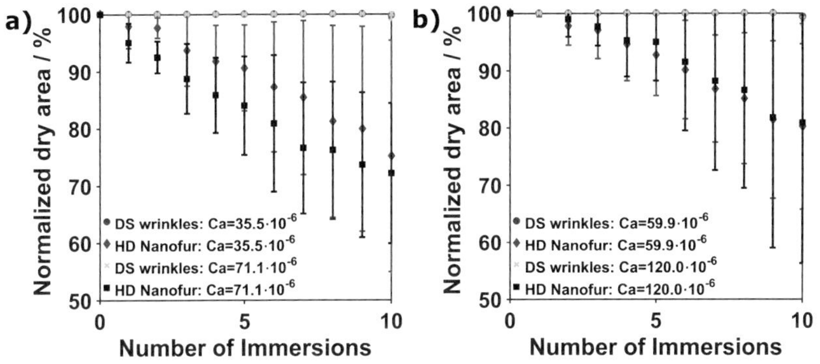

Figure 4.11: Contamination resistance of superhydrophobic surfaces. Averaged normalized dry area of DS wrinkles and HD nanofur as a function of number of immersions for different capillary numbers (Ca) corresponding to immersion in (a) 36 wt% and (b) 47 wt% glycerol/water solution at 50 mm·min^{-1} and 100 mm·min^{-1}.

During these experiments, superhydrophobic surfaces have been shown to exhibit anti-fouling properties, as long as the retained air layer is intact [160, 162–164]. Once the retained air layer collapses however, the increased surface area of superhydrophobic surfaces can increase cell adhesion, and, as a result promote bio-fouling [165, 166]. While most of the anti-fouling work to date has focused on stationary surfaces, an investigation into the dynamic stability of retained air layers against repeated immersion in liquids other than DI water, as well as investigating contamination resistance of superhydrophobic surfaces has been neglected. To measure contamination resistance and *in-situ* dynamic plastron stability, contact line motion measurements (Section 2.4.3) are combined with macroscopic air-water interface visualization (Section 2.4.4). This allows for precise tailoring of testing parameters in order to expose surfaces to application specific environments as described in Section 2.4.3.

To test resistance of HD nanofur and DS wrinkles against contamination and identify topographical features necessary for high contamination resistance, the two superhydrophobic surfaces were immersed in 36 wt% and 47 wt% glycerol/water solutions for up to ten times, while observing the evolution of the air-water interface. Figure 4.11 shows the average normalized dry area of DS wrinkles and HD nanofur as a function of number of immersions and capillary number (Ca). For all Ca the repeated immersion of HD nanofur results in a decay of the retained air layer, whereas the retained air layer on DS wrinkles stays 100% intact. Moreover, the decrease in dry area on HD nanofur, and thus contamination of the surface, increases with increasing immersion velocity. The dependency of plastron decay on immersion velocity is in agreement with observations on surfaces immersed in water (Fig. 4.9, Fig. 4.10), where increased immersion velocity also resulted in increased plastron decay. However, in contrast to immersion in glycerol, repeated immersion in DI water with a constant immersion velocity did not result in a further decrease of dry area. Surprisingly, an increase in glycerol concentration did not affect the decay rate of the retained air layer on HD nanofur as shown in Fig. 4.11 b).

To investigate if surface contamination is time dependent, HD nanofur was immersed in both prepared glycerol solutions for 24 hours. Comparison of the dry area before and after immersion revealed no significant differences, showing that contamination is not time dependent. Hence, contamination is a dynamic effect and occurs during immersion or movement of the sample. During immersion, the high repellency of the surface results in a continuously increasing curvature of the air-liquid interface, until the dynamic advancing CA is reached. The increase in curvature results in a constantly increasing Laplace pressure, which could be responsible for the observed wetting transition. An increase in immersion velocity additionally results in a faster increase in Laplace pressure, which, in turn destabilizes the retained air layer. Combined with the higher shear forces acting on the retained air layer with increasing immersion velocity, this results in a velocity dependent plastron decay rate. The contamination of nanofur was permanent and the films could not be cleaned by prolonged immersion in DI water or ethanol, or sonication in ethanol.

As mentioned before, the retained air layer on DS wrinkles is 100% intact for all Ca, thus, indicating a superior contamination resistance compared to HD nanofur (Fig. 4.11). This superior resistance is the result of several topographical features prevalent on DS wrinkles and missing on HD nanofur: (i) smaller length scale of topographical features (nm); (ii) threefold hierarchy ranging from nanometers to micrometers instead of double hierarchy of microcavities and nanohairs on nanofur; (iii) low surface energy resulting in minimal adhesion; (iv) microwrinkles forming reentrant structures. Comparing these resistance boosting features with the dynamic stability enhancing features from Section 4.2.1, reveals that the same topographical features responsible for high dynamic plastron stability are also responsible for a high contamination resistance.

4.2.4 Directionality of Dynamic Wetting Transitions

To investigate how wetting transitions spread on superhydrophobic surfaces, the dynamic CA of MD nanofur was measured as a function of immersion depth for repeated immersion in a 30 wt% sucrose/water solution at 20 mm min^{-1} as shown in Fig. 4.12 a), b). During the first immersion, the receding CA shows almost no change across the complete sample surface. However, after repeated immersions, the receding CA decreases linearly towards the lower end of the nanofur substrate. This linear decay becomes more pronounced with increasing immersion repetition and finally leads to a completely wetted and contaminated lower surface area after ten immersions, as shown in Fig. 4.12 a). The gradual decrease of receding CA with increasing number of immersions, indicates that a superhydrophobic surface passes through several intermediate states between intact air layer and completely contaminated hydrophilic surface, which is in contrast with the expected two distinct wetting states. Moreover, the change in wetting behavior is not as pronounced in the dynamic advancing CA as it is in the dynamic receding CA, as shown in Fig. 4.12 b). Repeated immersion of MD nanofur results in a lower mean value and higher scattering of the advancing CA, but does not display the same linear trend as can be observed for the receding CA. The reason for this is most likely the lower susceptibility of the advancing CA in regard to pinning of the liquid on the surface.

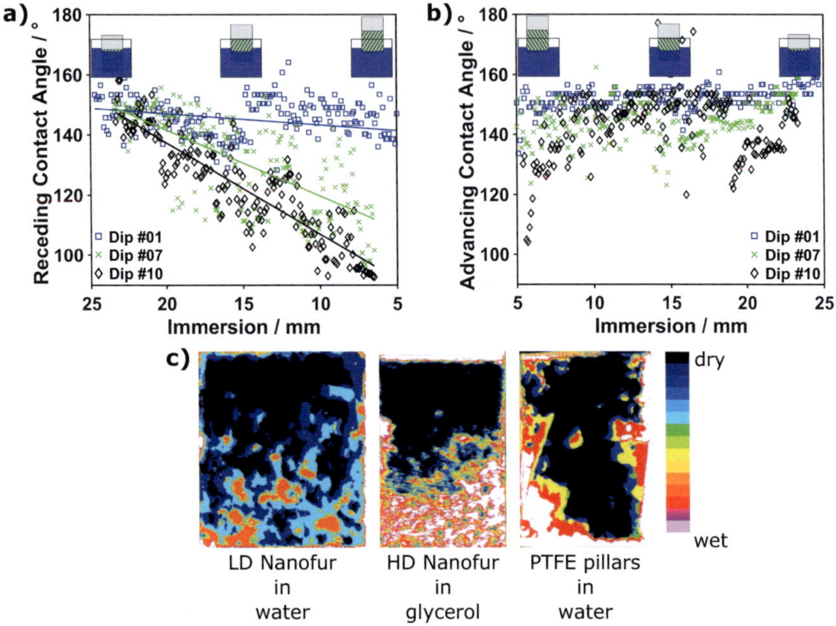

Figure 4.12: Directionality of dynamic wetting transitions. Dynamic (a) receding and (b) advancing CA of MD nanofur as a function of immersion depth for repetitive immersion in 30 wt% sucrose/water solution at 20 mm min^{-1}. Lines represent linear regression fit of the respective data. Inset displays position of nanofur at corresponding immersion depth. c) Overlaid images of intact air layer on superhydrophobic surfaces after immersion. LD nanofur and PTFE pillars were immersed in DI water at 20 mm min^{-1} - 200 mm min^{-1}. HD nanofur was immersed in 36 wt% glycerol solution at 50 mm min^{-1} for 10 times.

To further verify directionality of dynamic wetting transitions and contamination on superhydrophobic surfaces, photographs of the air-water interface on superhydrophobic surfaces during repeated immersions in water and glycerol were overlapped as shown in Fig. 4.12 c). Plastron images of LD nanofur and PTFE pillars after immersion in DI water at 20 mm min^{-1}, 50 mm min^{-1}, 100 mm min^{-1} and 200 mm min^{-1} were overlapped, while additionally HD nanofur images after each of 10 consecutive immersions in 36 wt% glycerol/water solution were overlapped. The merged images show a clear directionality, with wetting transitions starting at the lower end of the superhydrophobic surface and then spreading upwards. This is in agreement with the receding CA measurements shown in Fig. 4.12 a). The directionality is possibly due to the build up of the contact line during immersion in the liquid. During immersion, the superhydrophobic surface's high repellency results in a liquid curvature that is constantly increasing until the advanced dynamic CA is reached. This increase in curvature is equivalent to an increase in Laplace pressure, and combined with shear forces acting on the moving air-liquid-interface can push the liquid into the topography, resulting in a wetting transition. Because the liquid curvature changes the fastest at the edge of the superhydrophobic surface, this area also experiences the largest increase in Laplace pressure. Thus, the edge of the superhydrophobic film is the likeliest spot for a wetting transition to occur. Once a wetting transition/contamination occurs at the edge of the film, it moves the superhydrophobic/hydrophilic border upwards onto the structured area, where the same phenomenon is repeated, resulting in a gradual progression of the wetting transition upwards along the surface. Lastly, Fig. 4.12 c) shows that wetting transitions do not start in random spots, but rather spread outwards from already wetted areas. Once a wetting transition occured, the wetted area gradually spreads outward into all directions. As a result, the overall dynamic plastron stability is limited by the area with the weakest resistance against a wetting transition.

4.3 Conclusion

In conclusion, static and dynamic wetting transitions of bioinspired superhydrophobic nanofur and other selected superhydrophobic surfaces were characterized and the effect of surface energy and varying topographical features on the retained air layer stability was investigated. In case of elevated hydrostatic pressure, a combination of slightly hydrophilic material and superhydrophobic surface topography was shown to exhibit a higher plastron stability than the same topography with highly hydrophobic/hydrophilic bulk wetting properties. Utilizing the high reflectivitiy of the air-water interface, the macroscopic stability of the retained air layer on nanofur topography was studied and an exponential decrease of dry area with increasing hydrostatic pressure was demonstrated with only 50% dry area still intact at 0.5 bar. The experimental values were compared to theoretically predicted critical pressure and are in good agreement. Additionally, by using perforated nanofur and introducing an external pressure support, inspired by pressure-stable air layers on aquatic insects, the retained air layer on nanofur was stabilized and hydrostatic pressure limitations were overcome. By adjusting the retained air pressure to the hydrostatic pressure acting on the air-water interface, more than 86% of the retained air layer remains intact at hydrostatic pressures of more than 4 bar, which corresponds to 40 m water depth and a stability increase of several orders of magnitude. The pressure support further increases the time stability of the retained air layer to more than 4 hours at up to 3.5 bar. The stability enhancing mechanism was additionally shown to be independent of perforation parameters, such as number and position of pores. By increasing the retained air pressure above the hydrostatic pressure a buffer pressure, which stabilizes the retained air layer against pressure fluctuations of up to 3 bar, was introduced.

Furthermore, by combining high reflectivity and contact line motion measurements, the dynamic plastron stability of three different superhydrophobic surfaces displaying variations in structure regularity, hierarchy, bulk wetting properties and other topographical features was measured. Six topographical properties were identified as essential for a high dynamic stability of the retained air layer of superhydrophobic surfaces: (i) hierarchical topography; (ii) high density roughness; (iii) minimal water-solid-contact area; (iv) nanometer-scale feature size; (v) surfaces with cavities or cavity-like structures and (vi) reentrant structures. The same topographical properties are additionally shown to be beneficial for increasing resistance against contamination, but also reduce the surfaces drag-reducing capabilities. Moreover, wetting transitions and contamination are shown to spread from the bottom of a surface upwards and from broken down areas outward. Thus, the dynamic plastron stability is weakest at the lower edge of the superhydrophobic surface and limited by the area with the lowest stability against a wetting transition. Finally, by using contact line

motion measurements, surface contamination and the decay of wetting properties was shown to be a gradual process.
Because dynamic environments have mainly been ignored in respect to retained air layer stability but are essential for drag-reducing applications, further investigation into stability enhancing parameters can provide a basis for fabricating highly stable air retaining surfaces. Especially the ease and low complexity of the used measurements addresses the lack of appropriate characterization techniques. Additionally, the versatile possibilities to implement the presented stability enhancing mechanisms into other superhydrophobic surfaces widen the applicability of air-retaining surfaces for drag reduction. Lastly, characterisation of drag-reducing capabilities of pressure-supported perforated nanofur and adaptation of the stabilization mechanism to other drag-reducing mechanisms are promising future studies.

A shortened version of the retained air layer stability measurements on as prepared thin nanofur (Section 4.1.2) was published as part of the article "Bioinspired Air-Retaining Nanofur for Drag Reduction", Maryna N. Kavalenka, Felix Vüllers, Simone Lischker, Claudia Zeiger, Andreas Hopf, Michael Röhrig, Bastian E. Rapp, Matthias Worgull, and Hendrik Hölscher, ACS Applied Materials and Interfaces, 2015, 7, 10651-10655.

A shortened version of the laser confocal microscope study and the study of pressure supported perforated nanofur (Sections 4.1.1,4.1.3-4.1.4) was published in "Pressure-Stable Air-Retaining Nanostructured Surfaces Inspired by Natural Air Plastrons", Felix Vüllers, Yann Germain, Luce-Marie Petit, Hendrik Hölscher and Maryna N. Kavalenka, Advanced Materials Interfaces, 2018, 5, 1800125-1800132

The investigation into dynamic plastron stability (Section 4.2) will be published as "Dynamic Wetting Behavior and Plastron Stability on Superhydrophobic Surfaces", Felix Vüllers, Sam Peppou-Chapman, Maryna N. Kavalenka, Hendrik Hölscher and Chiara Neto

5 Superhydrophobic Surfaces for Optoelectronic Applications

Accumulation of natural and anthropogenic contamination on photovoltaic (PV) modules poses an immense problem, as it drastically decreases in-coupling of light into the active medium of the PV module, and, thus, decreases efficiency by up to 50% [34, 37, 39, 40]. Active cleaning mechanisms would mitigate those effects, but also result in a significant increase in operational costs [34–38, 41]. In contrast, minimal adhesion of particulate matter on superhydrophobic surfaces and the resulting self-cleaning properties present a long term passive solution, which drastically decreases maintenance costs. If additionally coupled with high transmission or anti-reflective behavior, the coatings can even increase the efficiency of optoelectronic devices, thus, making transparent superhydrophobic surfaces one of the most promising solutions for soiling [34]. Superhydrophobic thin nanofur is inspired by the dense layer of transparent microhairs present on the leaves of *Salvinia cucullata* and *Pistia stratiotes* (Fig. 3.3, Section 3.1.2), which renders the leaves superhydrophobic without impairing the absorption of light necessary for photosynthesis and the plants survival [167].

The first part of this chapter (Section 5.1) investigates the application of thin nanofur on optoelectronic devices. First the self-cleaning properties of thin nanofur are tested in Section 5.1.1, before its optical properties are characterized in Section 5.1.2, and its effect on the efficiency of optoelectronic devices is studied in Section 5.1.3. In the second part of the chapter (Section 5.2) a microcavity array based on nanofur topography and transferred to a PV relevant polymer (FEP) in order to fulfill application specific material constraints, is investigated as an efficiency enhancing, self-cleaning coating for PV modules.

5.1 Nanofur Application on Optoelectronic Devices

To reduce soiling and consequently efficiency losses of optoelectronic devices, surface coatings combining superhydrophobicity, self-cleaning properties and high optical transmission are necessary. To achieve this, thin nanofur was fabricated as described in Section 3.1.2 and classified as superhydrohobic through its extreme wetting properties (Section 3.4.1). Furthermore, thin nanofur is highly flexible, and, thus, can be applied to any surface geometry, making it widely applicable in light managing applications.

5.1.1 Self-Cleaning Properties of Nanofur

Most superhydrophobic nano- and microstructured surfaces exhibit self-cleaning properties known as "Lotus-Effect" [1, 11, 41, 168]. Because of the superhydrophobicity, water droplets roll down these structured surfaces at small tilt angles, pick up any contaminants they come in contact with, and, thus, clean the surface in the process. This self-cleaning behavior stems from the high roughness of the nano- and microstructured surface, which minimizes the contact area between contaminant and solid substrate (Fig. 5.1 b). The minimal contact area results in an adhesion between contaminant and substrate that is lower than the adhesion between water and contaminant, allowing the droplet to pick up the particle and carry it off [1, 168, 169]. On an unstructured surface, the greater contact area, and, thus, higher adhesion between contaminant and substrate results in water droplets creeping down the surface and simply rearranging the contaminants as shown in Fig. 5.1 a) or sticking to the surface until they evaporate [170].

Self-cleaning properties of thin nanofur were tested by contaminating unstructured PC as well as thin nanofur with graphite powder (particle size: 30 µm to 900 µm) and positioning the surfaces on a glass slide tilted at 30°, before dispensing single consecutive water droplets on both surfaces. On unstructured PC the water droplets stick to the surface without exhibiting any self-cleaning properties, whereas they roll down the thin nanofur film, picking up the graphite particles and cleaning the surface in the process (Fig. 5.1 c). To test a wide range of contaminant species and sizes, the self-cleaning experiment was repeated with silver powder, clay, aluminum oxide particles and paprika powder (Fig. 5.1 d). Dispensed water droplets cleaned almost all of the different contaminants off the surface, with the exception of particles which are significantly smaller than the diameter of the prevalent microcavities. These particles fell into the microcavities, where they could not be reached by the water droplets, and, thus, not get cleaned off.

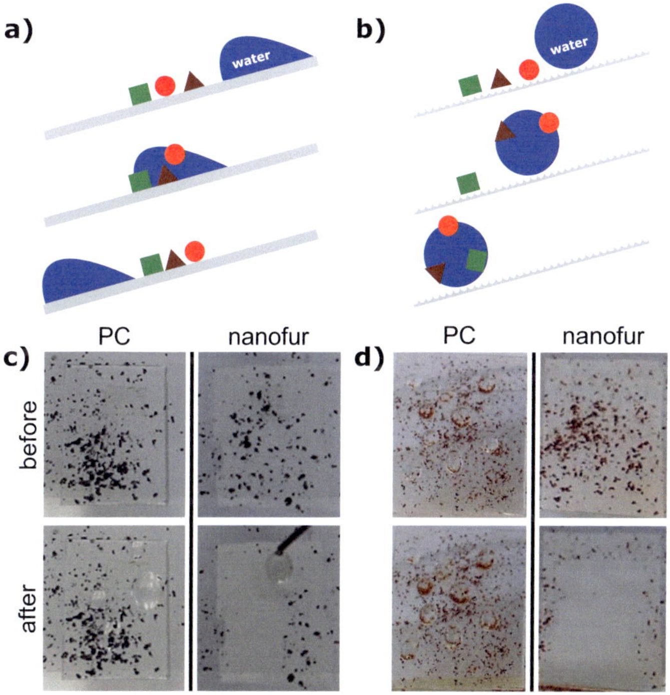

Figure 5.1: Self-cleaning properties of superhydrophobic thin nanofur. a), b) Schematic illustration of self-cleaning effect on micro- and nanostructured superhydrophobic surfaces. Adapted with permission of Barthlott *et al.* [1]. c), d) Self-cleaning properties of thin nanofur compared to unstructured PC. First, graphite powder (c) and paprika powder (d) were distributed on both surfaces, before dispesing consecutive water droplets. On unstructured PC water droplets stick to the surface, while they roll down the thin nanofur film, picking up contaminants and effectively cleaning the surface in the process. (c) is reprinted with permission from Vüllers *et al.* [72]. Copyright 1999-2018 John Wiley & Sons, Inc.)

5.1.2 Optical Properties of Nanofur

Besides self-cleaning properties, application of thin nanofur on optoelectronic devices requires the films to exhibit excellent optical properties. To ensure high transmittance, thin nanofur was fabricated from optical grade PC with a refractive index of $n = 1.58$. Furthermore, as mentioned in Section 3.1.2 the fabrication procedure is adapted to minimize roughness on the backside of thin nanofur, and, thus, reduce possible reflection losses due to backscattering.

The optical properties of thin nanofur were characterized by measuring the overall reflection using an UV/VIS/NIR-spectrophotometer with an integrating sphere as described in Section 2.4.6. For these measurement thin nanofur films were attached to a black PC sheet absorbing 95% of light in a spectral range of 300 nm to 800 nm. To minimize errors during the measurement, a reference reflection measurement of the black PC plate was taken and all measured reflection profiles were corrected accordingly. Figure 5.2 a) shows reflection profiles of as prepared thin nanofur (dry nanofur) and thin nanofur with its backside wetted with DI water (wetted nanofur) compared to the reflection profiles of unstructured PC with (wetted PC) and without (dry PC) a wetted backside. Used as a transparent coating, the as-prepared thin nanofur shows moderate reflection (13.7% at $\lambda = 550$ nm) for wavelengths ranging from 300 nm to 800 nm, comparable to the reflection of unstructured PC (10.4% at $\lambda = 550$ nm). However, wetting the backside of thin nanofur reduces the reflection for visible light to less than 4% (1.6% at $\lambda = 550$ nm), which is slightly lower than the reflection of unstructured wetted PC (2% at $\lambda = 550$ nm) used for commercial applications. The reduced reflection of wetted nanofur compared to dry nanofur is due to the water substituting the air layer between the substrate and the attached nanofur. The intermediate refractive index of water ($n = 1.33$) bridges the refractive index gap between the two surfaces, which in turn reduces reflection, as described by Fresnel's law [171]. The high transparency of thin nanofur is qualitatively demonstrated by attaching dry and wetted nanofur to a printed laminated paper as shown in Fig. 5.2 b). From a top view, the writing can clearly be seen through both nanofur films. However, from a tilted view, glare on the dry nanofur obscures the writing, while it can still be clearly seen through the wetted nanofur, thus, highlighting the broadband reduction in reflection of wetted nanofur compared to dry nanofur.

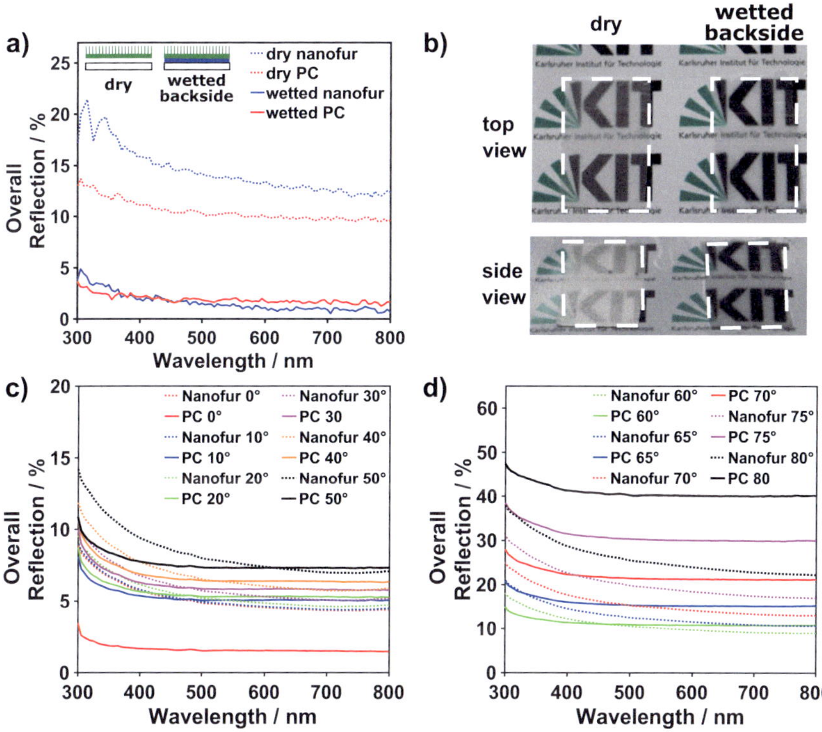

Figure 5.2: Reflectivity of thin nanofur. a) Typical reflection spectra of as-prepared thin nanofur (dry nanofur), thin nanofur with a wetted backside (wetted nanofur) and unstructured PC with (wetted PC) and without (dry PC) wetted backside. Inset shows schematic attachment of thin nanofur to black polymer. b) Top and side view images of laminated paper coated with thin nanofur. Writing is clearly visible through wetted nanofur for top and side view, indicating a reduced reflectivity compared to as-prepared thin nanofur. c), d) Overall reflection of thin nanofur and unstructured PC as a function of wavelength and AOI. Nanofur shows reduced reflection for wavelengths greater than 500 nm and for high AOI. (a), b) are reprinted with permission from Vüllers *et al.* [72]. Copyright 1999-2018 John Wiley & Sons, Inc.)

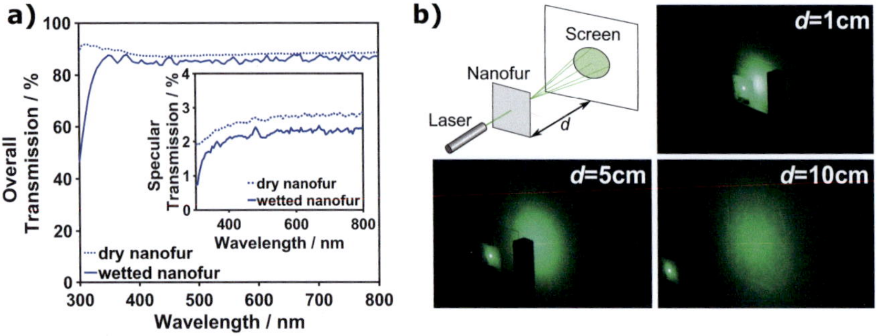

Figure 5.3: Transmission of thin nanofur. a) Overall transmission spectra of as-prepared self standing thin nanofur film (dry nanofur) and thin nanofur film attached to glass slide by water (wetted nanofur). The inset shows the respective specular transmissions. b) Light of a laser pointer transmitted through thin nanofur and observed in varying nanofur-screen distances (*d*). Widening of the beam and the low specular transmission demonstrate the high forward scattering properties of thin nanofur topography. (a) is reprinted with permission from Vüllers *et al.* [72]. Copyright 1999-2018 John Wiley & Sons, Inc.)

Additionally, the angle resolved overall reflectivity of thin nanofur from a normal angle of incidence (AOI) of 0° to a glancing AOI of 80° is compared to that of unstructured PC in Fig. 5.2 c), d). For low AOI, nanofur and unstructured PC show similar reflectivity, with PC having lower reflection values for shorter wavelengths (< 500 nm) (Fig. 5.2 c). Surprisingly, with increasing AOI, the nanofur reflectivity first increases faster than that of unstructured PC up to 50° tilt, where the contrast between thin nanofur and PC is minimal for high wavelengths and maximum for low wavelengths. For AOI $> 50°$ nanofur then significantly reduces the reflection to almost half the value of unstructured PC (for high wavelengths), making it most valuable for optoelectronic devices installed at increased tilt angles. The reduced reflectivity of thin nanofur for wavelengths ranging from 550 nm to 750 nm for almost all AOI is especially beneficial for PV modules, as the photon flux of the sun is highest in this spectral range, and, thus, a reduced reflectivity results in a significant boost in efficiency.

Besides attachment of thin nanofur as a transparent and self-cleaning coating, it can also be used as a translucent self-standing superhydrophobic film with high transmission in the visible spectrum (300 nm to 800 nm). Figure 5.3 a) shows the typical transmission spectra of a self-standing thin nanofur film and a thin nanofur film with wetted backside attached to a glass slide. More than 88% of light inciding on the dry nanofur is transmitted through the film, while wetting the backside of the nanofur and attaching it to a glass slide (wetted nanfour) still shows transmission values above 85% for the visible spectrum. The reduced transmission of wetted nanofur is due to light absorption of both water and glass [172]. Furthermore, the low specular transmission of thin nanofur (inset of Fig. 5.3 a) indicates transmitted light to be highly diffusive. For wavelengths ranging from 300 nm to 800 nm less than 2.8% of inciding light transmits thin nanofur without being scattered (2.6% for dry and 2.5% for wetted nanofur at 550 nm). Thus, the haze factor of thin nanofur, defined as ratio of diffusive transmitted light to overall transmitted light, is greater than 97% proving the high amount of forward scattering. The high forward scattering is further qualitatively demonstrated in Fig. 5.3 b) where the light of a laser pointer transmitted through thin nanofur is observed on a white screen positioned at varying distances. For small nanofur-screen distances (d) the light hardly spreads on the screen. However, with increasing distance, the illuminated area widens further and further, thus, demonstrating the change of direction of transmitted light. High forward scattering of light stems from the nanofur surface topography consisting of microcavities surrounded by a dense layer of micro- and nanohairs. This unique surface topography randomizes the propagation angle of transmitted light due to the random orientation of the micro- and nanohairs. In combination with the microcavities acting as microlenses, the randomized propagation results in the high amount of forward scattering, and, thus, the translucent properties of thin nanofur.

5.1.3 Application of Nanofur on Optoelectronic Devices

The exceptional optical properties of thin nanofur demonstrated in the previous section can be exploited for light managing applications in optoelectronic devices, such as light extraction in organic light emitting diodes (OLED) or light incoupling in PV modules. The organic light emitting diode (OLED) and EQE measurements presented in the following subsection were jointly carried out with Jan B. Preinfalk (Light Technology Institute (LTI), KIT) and Efthymios Klampaftis (Institute of Microstructure Technology (IMT), KIT).

Before applying thin nanofur to an OLED as a coating, the emission profile of light transmitting through the thin nanofur film was measured as described by Hünig *et al.* [173]. Thin nanofur was attached to a glass hemisphere using refractive index matching liquid and illuminated with a white LED and a beam collimator. The device was then mounted on a rotational stage and the angle resolved emission profile was acquired using an optical fiber mounted on a monochromator. Fig. 5.4 a) shows, that the emission profile is similar to the ideal lambertian profile. Therefore, a thin nanofur coated optical device has the same apparent brightness independent of viewing angle, which is an essential property for most lighting applications [174].

In order to investigate the effect of thin nanofur on OLED performance, thin nanofur films and unstructured PC were attached to monochromatic yellow-emitting OLEDs by dispensing the UV-curing adhesive Norland Optical Adhesive 86 on the OLED and pressing the respective films backside into it. Next the adhesive was crosslinked with a mercury lamp ($\lambda = 365$ nm) for 30 s. A schematic of the tested OLED as well as a photograph of an OLED half covered with nanofur is shown in Fig. 5.4 b). Compared to the uncoated half of the OLED, attachment of highly scattering thin nanofur widens the pixels of the OLED and seemingly increases the active area, and, as a result the OLED efficiency. The positive effect a thin nanofur coating has on OLED efficiency is verified by measuring the current and power efficiency of a bare monochromatic OLED and comparing it to the efficiency of OLEDs coated with thin nanofur and coated with unstructured PC, as shown in Fig. 5.4 c), d). Attachment of thin nanofur results in a more than 10% relative intensity enhancement of the coated OLED compared to a bare device, while attachment of unstructured PC has no effect on the average intensity. Without attached nanofur, photons generated by the OLED can be trapped inside the glass substrate due to the high contrast in refractive index resulting in total

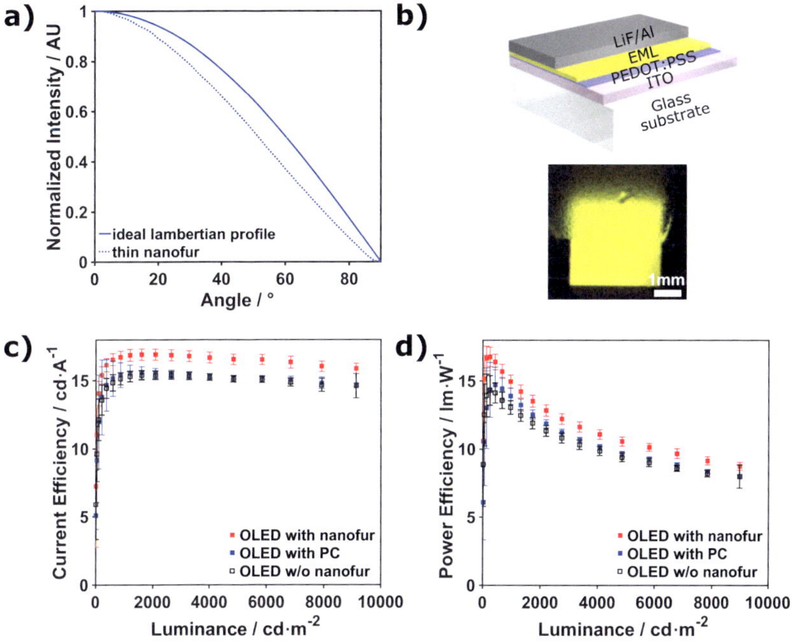

Figure 5.4: Effect of thin nanofur on OLED efficiency. a) Angle resolved emission profile of light transmitting a thin nanofur film compared to an ideal lambertian emission profile. b) Schematic of fabricated monochromatic yellow OLED and photograph of operating monochromatic OLED half coated with thin nanofur. c) Current efficiency as a function of luminance of a bare yellow monochromatic OLED, an OLED coated with thin nanofur and an OLED coated with unstructured PC. d) Power efficiency as a function of luminance of a bare yellow monochromatic OLED, an OLED coated with thin nanofur and an OLED coated with unstructured PC. Attachment of thin nanofur results in an average intensity enhancement of (c) 11% and (d) 14%, while the attachment of unstructured PC shows no enhancement. (Reprinted with permission from Vüllers *et al.* [72]. Copyright 1999-2018 John Wiley & Sons, Inc.)

internal reflection at the substrate-air interface [175, 176]. The presence of the highly scattering rough nanofur surface widens the escape cone, thus, suppressing substrate-guided modes and enhancing the out-coupling probability of generated photons, which increases the out-coupling efficiency [176, 177]. Even though, the efficiency enhancing effect of nanofur is only demonstrated for a yellow emitting OLED, the wavelength independent optical properties of nanofur (Fig. 5.2 a),

Fig. 5.3 a)) make the extension of these results to other emitters operating over the visible range possible.
Finally, the effect of thin nanofur on the light in-coupling properties of mc-Si solar cells was tested. To this end, the external quantum efficiency (EQE) of a bare, nanofur coated and unstructured PC coated solar cell were measured as described in Section 2.4.7 and are displayed in Fig. 5.5 a). To account for the multicrystalline nature of the solar cell and the random and highly heterogenous topography of the thin nanofur, five different spots were measured and averaged for each displayed configuration. The steep drop in EQE for the solar cell with attached PC and thin nanofur in the UV spectral region is due to high absorption of the refractive index matching liquid used for attaching PC and nanofur to the solar cell. The respective short-circuit current densitys are shown in Fig. 5.5 b) and were calculated based on the EQE measurements and by using Eq. (2.11). Attaching thin nanofur to the mc-Si solar cell resulted in a relative enhancement of 5.8% in J_{SC} compared to the bare device. Including measurements with unstructured PC allows to distinguish between the improvement caused by the introduction of PC and the improvement caused by the topographical structure of thin nanofur. Introducing unstructured PC increases the EQE throughout the whole spectral region relevant for generating power as shown in Fig. 5.5 a). However, for wavelengths ranging from 550 nm to 750 nm, where the photon flux of the sun is highest, the enhancement is less pronounced, because the prevalent anti-reflective coating of the utilized solar cell has been optimized for this region. The enhancement in EQE results in an increase of 1.1 mA cm^{-2}, which corresponds to a relative enhancement of 3.4% and is a result of PC (refractive index $n = 1.58$) reducing the gap in refractive index between air ($n = 1$) and the anti-reflective coating (a-SiN$_X$:H, $n \approx 2$ for λ =300 nm - 800 nm) of the solar cell. A similar effect occurs during glass encapsulation of PV modules for commercial use. Nonetheless, attachment of thin nanofur further increases the EQE for wavelengths ranging from 300 nm to 1100 nm, thus, increasing the short-circuit current density by an additional 0.8 mA cm^{-2} (2.4%) compared to unstructured PC. As the introduction of PC is already factored

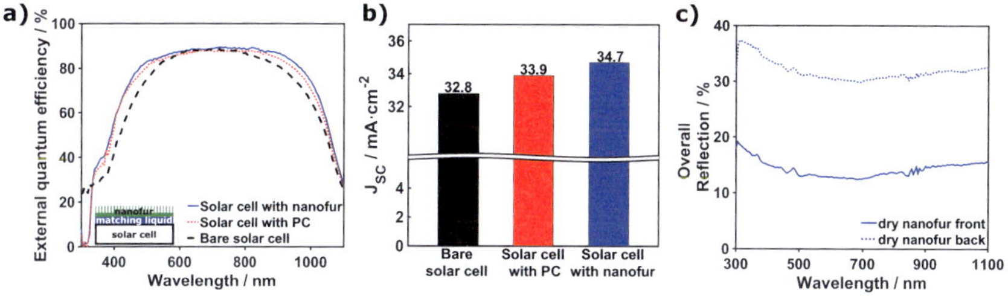

Figure 5.5: Effect of thin nanofur on mc-Si solar cells. a) EQE as a function of wavelength for a bare mc-Si solar cell, a solar cell with attached nanofur and a solar cell with attached unstructured PC under normal incidence. An average of five measurements is shown to account for discrepancies due to the multicrystallinity of the cell b) J_{SC} of the bare solar cell, the solar cell with attached nanofur and the solar cell with attached unstructured PC. Attachment of nanofur significantly increases EQE and J_{SC}. c) Typical reflection spectra of thin nanofur illuminated from the front and illuminated from the back. Reflectance from the back side is almost twice as high as from the front side. (a) is reprinted with permission from Vüllers *et al.* [72]. Copyright 1999-2018 John Wiley & Sons, Inc.)

out of this enhancement, it is solely due to the micro- and nanostructure of the thin nanofur film.

Besides efficiently transmitting and in-coupling light, the increased efficiency is partially the result of the difference in reflectance behavior for light inciding on thin nanofur from the structured (front) side and from the smooth (back) side (Fig. 5.5 c). Light hitting the PV device will be partially reflected at the nanofur-PV interface, and, thus, not enter the active medium and be lost for energy conversion. Therefore, to have an efficient light-harvesting coating, not only does the coating have to have good light in-coupling properties, but it also needs to suppress out-coupling of light reflected at the coating-substrate interface. This combination of high in-coupling and low out-coupling can be achieved through a material with low reflection when illuminated from the front, while simultaneously having a highly reflective backside. In this case, instead of being lost, light reflected at the PV interface is reflected back towards the photoactive layer, increasing absorption probability, and, as a result the electrical output. To investigate this behavior, the total reflection of thin nanofur illuminated from the front and back side was measured as a function of wavelength and is shown in Fig. 5.5 c). While the reflectance from the front side is low to moderate as mentioned before, the reflectivity of the backside is twice as high over the complete PV relevant spectrum. Thus, nanofur suppresses back scattering, while simultaneously reducing front side reflection essential for PV modules.

5.2 Self-Cleaning Microcavity Array for Photovoltaic Modules

The previous section demonstrated the excellent optical and self-cleaning properties of superhydrophobic thin nanofur. However, due to the fragile nature of the high aspect ratio micro- and nanohairs, thin nanofur has limited mechanical resistance and can easily be damaged by scratching (Section 4.1.3, Fig. 4.7) [50, 51]. Moreover, even though thin nanofur is fabricated from optical grade PC, which is commonly used for optical applications [79], elongated exposure to UV-irradiation changes the polymers chemical structure, resulting in a change in color from transparent to yellow, thus, limiting its outdoor applications [178]. Even though this change in chemical structure does not influence wetting and self-cleaning properties, it mitigates the long term passive applicability of PC coatings for enhancing PV modules. In order to overcome these shortcomings, the thin nanofur topography is deconstructed and its parts are separately optically characterized (Section 5.2.1). The optically relevant microcavity array (MCA) was then transferred into a fluorinated polymer, as described in Section 3.2 and in turn characterized in regard to its self-cleaning (Section 5.2.2) and PV enhancing properties (Section 5.2.3). Simulations and EQE measurements presented in the following section were jointly carried out with Benjamin Fritz (Light Technology Insitute (LTI), KIT) and Aiman Roslizar (Institute of Microstructure Technology (IMT), KIT).

5.2.1 Optical Properties of MCA

Thin nanofur topography consists of a dense layer of high aspect ratio micro- and nanohairs supported by an underlying disordered MCA as shown in Fig. 5.6 a).

To distinguish between the impact of the dense layer of hairs and the impact of the underlying MCA on the optical properties of thin nanofur, most of the micro- and nanohairs were removed in a dry etching step, with the exception of hairs with a large base diameter. As a result, thc underlying MCA was exposed as shown in Fig. 5.6 b). The total reflection of as-prepared nanofur and nanofur with exposed μ-cavities as a function of wavelength under normal incidence is shown in Fig. 5.6 c). Removal of the hairs reduces reflection of the film by 2% compared to thin nanofur, indicating that the micro- and nanohairs mitigate the anti-reflective properties of the underlying MCA.

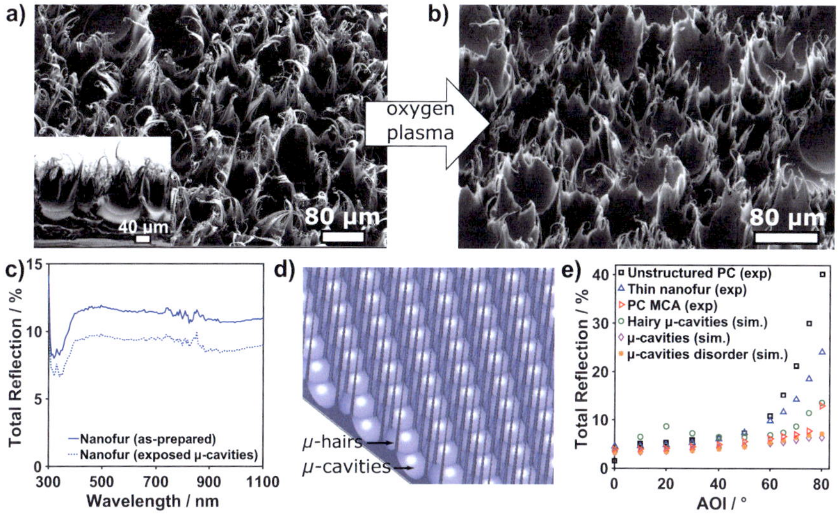

Figure 5.6: Experimental and numerical characterization of optical properties of thin nanofur. a) Top view and crossectional SEM image of thin nanofur. A dense layer of high aspect ratio micro- and nanohairs is supported by a disordered MCA. b) SEM image of exposed μ-cavities after oxygen plasma treatment removes most micro- and nanohairs. c) Total reflection of as-prepared nanofur and oxgen plasma treated nanofur under normal incidence. Removal of hairs reduces reflection lossed by 2%. d) Three-dimensional simulation model of MCA with included microhairs. Microhairs are simulated as 100 μm tall cones and positioned at corners of overlapping cavities. e) Experimentally obtained overall reflection of unstructured PC, thin nanofur and PC MCA and simulated overall reflection of PC MCA with and without microhairs and with and without disordered AR as a function of AOI at 600 nm. (Adapted with permission from Vüllers *et al.* [120]. Copyright 2018 American Chemical Society.)

To study the antireflective properties of the disordered µ-cavity topography and further investigate the effect of microhairs on the reflectivity of MCA films under varying AOI, a simplified MCA model with the possibility to include microhairs into the topography was developed (Fig. 5.6 d). Based on this model, all following simulations were undertaken. Because the µ-cavities on thin nanofur are more than one order of magnitude larger than the spectral range relevant for PV applications (300 nm - 1100 nm), optical properties of MCA are solely governed by ray optics and consequently all corresponding simulations are carried out using ray tracing simulations. A negative PDMS imprint (Fig. 3.9) shows that thin nanofur is covered in overlapping µ-cavities, which leave no planar surface area left. Therefore, the modelled cavities with radius r, were aligned in a hexagonal array with a lattice constant $a = \sqrt{3}r$, to achieve a complete tiling of the base. Moreover, all modeled µ-cavities were assigned an aspect ratio of $\mathrm{AR} = 0.45$, which was experimentally derived from cross-sectional SEM images of thin nanofur (Fig. 5.7). Numerical reflectivity simulations of MCA as a function of AR have shown minimal reflectivity for $\mathrm{AR} = 0.5$ (not shown here), indicating that the AR of thin nanofur is close to optimal, and, thus, making nanofur topography a promising template for anti-reflective coatings.

Experimental reflection values of thin nanofur, unstructured PC and PC MCA are compared to simulated reflection values of PC MCA with and without disorder and microhairs as a function of AOI in Fig. 5.6 e). The displayed reflection values are taken at $\lambda = 600\,\mathrm{nm}$, in order to be close to the solar irradiance peak. For unstructured PC and as-prepared thin nanofur reflection increases exponentially with increasing AOI up to 40% and 25% respectively. A similar exponential increase can be observed for the simulated MCA with incorporated microhairs. The increase in reflectivity is directly correlated to the height of the microhairs, as a reduction of cone height from 100 µm to 50 µm reduces reflectivity from 13.5% to 7.5%. The exponential increase in reflectivity for thin nanofur and MCA with microhairs is a result of the change in effective cross-sectional area of microhairs with increasing AOI. At normal incidence the area relevant for reflection is the horizontal cross-sectional area of the hairs, which for hairs with

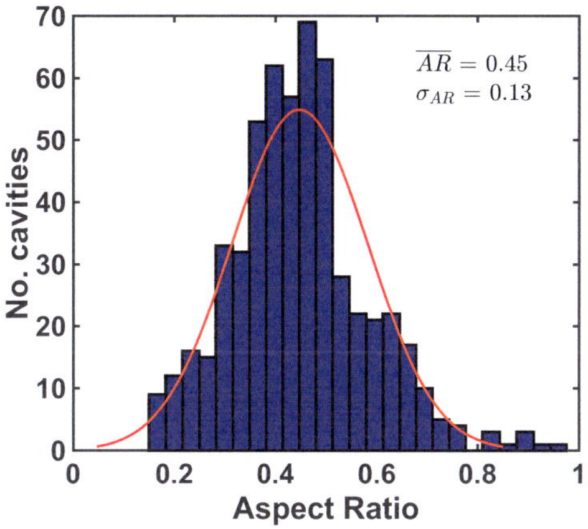

Figure 5.7: Experimentally derived AR of thin nanofur μ-cavities. AR follows a normal distribution with average $\overline{AR} = 0.45$ and standard deviation $\sigma_{AR} = 0.13$.

a diameter down to 200 nm is minimal compared to the bulk area. Therefore, incoming light is hardly influenced by the microhairs before interacting with the underlying MCA, and, as a result reflectivity is almost identical for a MCA with and without microhairs. However, with increasing AOI the greatly increased vertical cross-sectional area of the hairs becomes the effective area responsible for reflection. This increased area is more similar to an unstructured polymer, and, hence, increases reflection in the same way. Moreover, light hitting nanofur at high AOI encounters several layers of microhairs, and, thus, has to transmit through a higher number of refractive interfaces than for low AOI. At each refractive interface a small portion of light is reflected, and, as a result, an increase in number of refractive interfaces increases the total reflection. In contrast to surfaces with microhairs, simulated MCA without microhairs exhibit reflection values below 7% for all AOI. The influence of disorder in μ-cavity AR on the reflectivity of MCA films was investigated by fixing the modelled cavity diameter and varying the height of the individual cavities, thus, introducing disorder into the AR. To assign an AR to each μ-cavity of the modelled MCA, first the AR distribution of thin nanofur was measured using cross-sectional SEM images. Figure 5.7 shows that the AR of nanofur µ-cavities follows a normal distribution with mean value $\overline{AR} = 0.45$ and standard deviation $\sigma_{AR} = 0.13$.

Based on this distribution, modeled µ-cavities were assigned an AR. The introduced disorder changes the reflectivity of MCA films by less than 0.2% as shown in Fig. 5.6 e). This minimal reduction is in agreement with results shown by Fritz *et al.* [179] for microstructured surfaces composed of microcones. Hence, the effect of prevalent structural disorder on the surfaces optical properties can be neglected and, therefore, was not adjusted in the replication of MCAs (Section 3.2). Therefore, it can be concluded, that the increase in reflection of thin nanofur with increasing AOI compared to MCA without microhairs is solely a result of the microhairs. Thus, the bare MCA is more advantageous for applications requiring high transmittance, such as PV modules.

However, to achieve superhydrophobicity and self-cleaning properties with high surface energy materials such as PC, the dense layer of microhairs is essential, because it reduces the contact area between water and substrate necessary to achieve a Cassie-Baxter wetting state [54, 61]. Therefore, in order for low AR structures such as MCA to achieve superhydrophobicity and self-cleaning properties, while keeping good optical properties, a polymer with low surface energy, such as a fluorinated polymer has to be used. To combine optimal optical and wetting properties, the MCA topography was transferred into fluorinated ethylene propylene (FEP) using hot embossing as described in Section 3.2. A further benefit of using FEP is its high chemical and mechanical resistance against impact and tearing.

5.2.2 Self-Cleaning properties of FEP MCA

As mentioned in Section 5.1.1, many micro- and nanostructured superhydrophobic surfaces exhibit self-cleaning properties known as "Lotus-Effect" [1, 11, 41, 168]. This effect is especially of interest for photovoltaic (PV) devices as a way to reduce soiling, and, thus, maintenance and operational costs [35, 38, 40]. To maximize efficiency of PV modules, they are installed at different tilt angles depending on several constraints including geographical location, fraction of diffuse vs. direct sunlight and local shading from surrounding hills, trees and buildings [180, 181]. Therefore, in order to include optimal PV installation angles for most of the world [180], self-cleaning properties of FEP MCA were tested for tilt angles ranging from 10° to 45°. Furthermore, to encompass a variety of contaminants, sea sand (particle size: 100 μm - 315 μm) and graphite powder (particle size: 30 μm - 900 μm) were used as contamination. The large size distribution of graphite powder additionally tests the self-cleaning properties for a broad range of contaminant sizes.

After contaminating FEP MCA and unstructured FEP with graphite powder, consecutive water droplets were dispensed on both surfaces. The self-cleaning properties of FEP MCA and unstructured FEP at 10° and 45° are compared in Fig. 5.8 a), b). For all tilt angles, water droplets stuck to unstructured FEP resulting in no self-cleaning effect. Water droplets dispensed on FEP MCA on the other hand, rolled down the surface, picking up contaminants and effectively cleaning the surface in the process. Furthermore, optical microscope pictures of unstructured FEP and FEP MCA after conducting the self-cleaning test show graphite particles still contaminating unstructured FEP, while FEP MCA is completely cleaned from all contaminants (Fig. 5.8 a), b), c)).

Because sand and dust are the most common contaminants in arid regions, where installation of PV modules, and, therefore soiling of PV modules is of great interest, self-cleaning tests were repeated with sea sand. FEP MCA exhibited self-cleaning behavior for all tested angles, whereas sea sand and water droplets stuck to unstructured FEP, resulting in no self-cleaning behavior and a contaminated surface as shown in Fig. 5.8 d).

One main natural source providing water for self-cleaning properties of PV modules is rain. However, rain impacting on a superhydrophobic surface can result in a Cassie-Baxter to Wenzel wetting transition [182–185]. Such a transition would result in droplets sticking to the surface and a loss of superhydrophobicity and self-cleaning properties. Additionally, if droplets from dust rain stick to the surface of

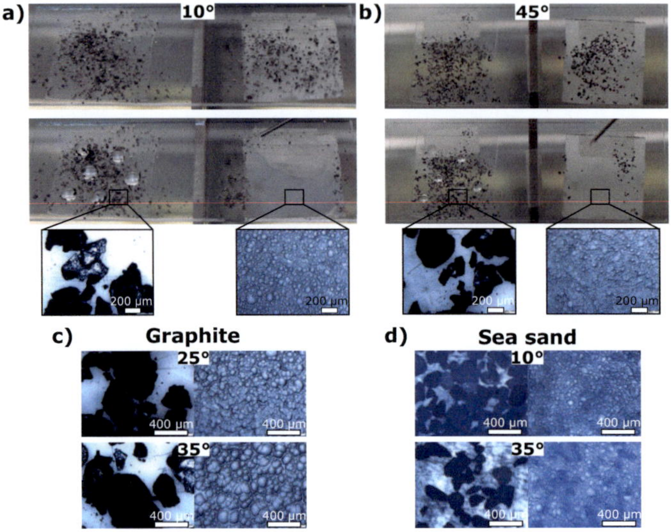

Figure 5.8: Self-cleaning properties of fluorinated ethylene propylene (FEP) microcavity arrays (MCA). Graphite is dispensed on FEP MCA (right) and unstructured FEP (left) tilted at (a) 10° and (b) 45° before single water droplets are dispensed on both surfaces. Independently of tilt angle, droplets stick to unstructured FEP, while they roll down the FEP MCA and pick up contaminants, thus, effectively cleaning the surface. Magnifications show contaminated unstructured FEP and cleaned FEP MCA after conducting the self-cleaning test. c) Microscope images of contaminated unstructured FEP (left) and cleaned FEP MCA (right) tilted at 25° and 35° after self-cleaning procedure with graphite powder. d) Microscope images of contaminated unstructured FEP (left) and cleaned FEP MCA (right) tilted at 10° and 35° after self-cleaning procedure with sea sand. (a), c), d) are reprinted with permission from Vüllers *et al.* [120]. Copyright 2018 American Chemical Society.)

PV modules, they can deposit particulate matter, further contaminating the module and reducing electrical output [186, 187]. In order to prevent this contamination and exhibit passive, rain induced self-cleaning properties, the wetting behavior of FEP MCA has to be highly stable against water droplets impacting under various speeds. To test the stabilty of the surface and its wetting properties against impacting water droplets, 13 µl droplets of DI water were dropped from 5 cm and 10 cm height onto a FEP MCA and an unstructured FEP surface tilted at 0°, 10°, 20° and 30°. The drop heights correspond to drop velocities of $1.08 \pm 0.05\,\mathrm{m\,s^{-1}}$ and

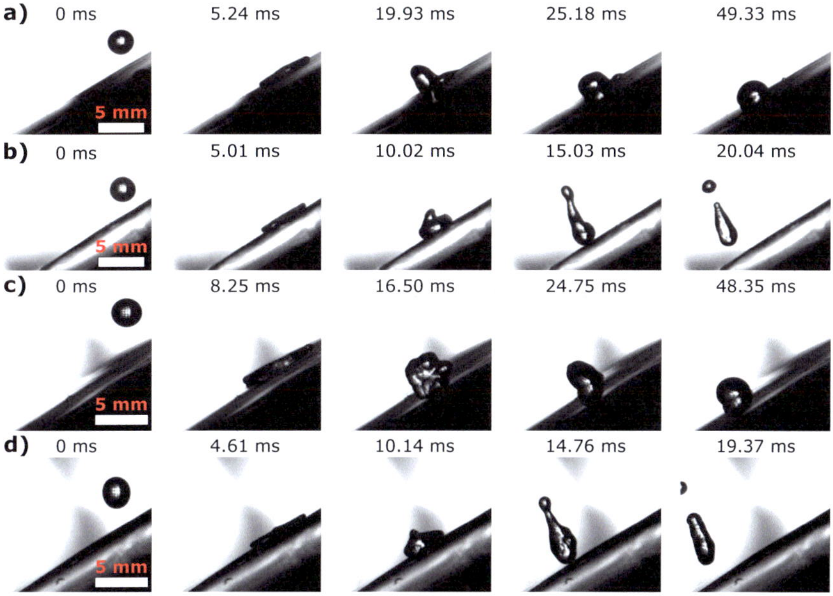

Figure 5.9: Bouncing droplets on FEP MCA. Time lapse series of a 13 µl DI water droplet impacting on (a, c) unstructured FEP and (b, d) FEP MCA at (a, b) $1.08 \pm 0.05\,\mathrm{m\,s^{-1}}$ and (c, d) $1.38 \pm 0.04\,\mathrm{m\,s^{-1}}$. For both velocities, water sticks to unstructured FEP and bounces off FEP MCA, demonstrating the high stability of wetting properties on FEP MCA. (a), b) reprinted with permission from Vüllers *et al.* [120]. Copyright 2018 American Chemical Society.)

$1.38 \pm 0.04\,\mathrm{m\,s^{-1}}$, and are in the same order of magnitude as rain drop velocities, which range from $2\,\mathrm{m\,s^{-1}}$ to $9\,\mathrm{m\,s^{-1}}$ [188]. Impacting droplets were recorded with a high speed camera and time-lapse image series of droplets impacting unstructured FEP and FEP MCA are displayed in Fig. 5.9. A droplet impacting on unstructured FEP deforms into a pancake and wets the complete covered surface area, before elongating into an ellipsoid and then recovering its hemispherical shape. During the whole deformation, the droplet is in contact with the surface. Because of the inherent hydrophobicity of FEP and the increased tilt, the droplet slides a short distance down the unstructured surface, but quickly comes to a complete halt and firmly sticks to the surface. On FEP MCA the impacting droplet also first deforms into an ellipsoid. However, due to superhydrophobicity of the FEP MCA film, the droplet does not wet the surface and reforms into a sphere, before its own kinetic energy propels it away from the surface, while forming satellite droplets. On further impact the droplet deforms and bounces off the surface again, until enough kinetic energy is transferred for the droplet to rest on the surface and simply roll down, possibly picking up contaminants on the way. An increase in velocity of droplets impacting unstructured FEP results in a longer creep distance but the droplets still come to a complete halt, whereas droplets bounce off FEP MCA films independent of tilt angle and drop velocity.

Therefore, FEP MCA films exhibit stable wetting properties under impact and contamination, and, thus, are superior to unstructured FEP films for PV application. Stable superhydrophobicity can furthermore keep the top cover of PV modules dry, thus, reducing surface conductivity, which is beneficial, because surface conductivity is known to cause potential-induced degradation of PV modules. This degradation can in turn reduce the electrical output or even cause failure of PV modules [189].

5.2.3 Application of FEP MCA on PV Modules

The effect of FEP MCA on the electrical output of PV modules was investigated by optically coupling FEP MCA films to glass-encapsulated multicrystalline silicon (mc-Si) solar cells. Similar to Section 5.1.3 all optical characterizations were additionally undertaken with unstructured FEP in order to differentiate between reduced Fresnel reflection due to the intermediate refractive index of introduced FEP ($n = 1.34$) and reduced reflectivity due to the microstructure of FEP MCA. Figure 5.10 a) compares the current-voltage (I-V) curves of a bare PV device, and a PV device with unstructured FEP and FEP MCA films optically coupled to the device. While attachment of unstructured FEP to the PV module results in a 3.7% relative gain in short-circuit current (I_{SC}) compared to the bare device, attaching FEP MCA increases the short-circuit current by a total of 4.6%. Therefore, the 3.7% increase is due to the intermediate refractive index of FEP, while

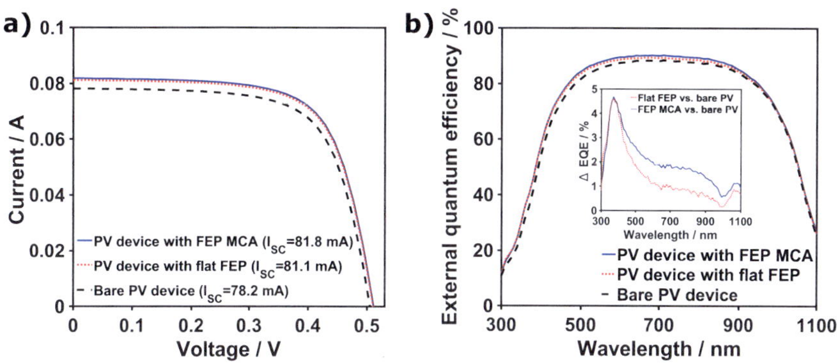

Figure 5.10: Effect of FEP MCA on PV devices. a) I-V curves of bare, unstructured FEP coated and FEP MCA coated glass-encapsulated mc-Si solar cell. b) EQE as a function of wavelength for bare, unstructured FEP coated and FEP MCA coated glass-encapsulated mc-Si solar cell under simulated sunlight (air-mass 1.5 global spectrum) and normal incidence. Compared to the bare device, unstructured FEP and FEP MCA increase the electrical gain by 3.7% and 4.6% respectively. Inset shows highest improvement of electrical output in the visible spectrum (λ = 400 nm - 700 nm). (Reprinted with permission from Vüllers *et al.* [120]. Copyright 2018 American Chemical Society.)

the additional 0.9% increase of FEP MCA is solely due to the microstructure of the embossed film. The optical improvement of FEP and FEP MCA is verified by EQE measurements shown in Fig. 5.10 b). Optically coupling FEP and FEP MCA to the PV device increases the EQE over the complete PV relevant spectrum (300 nm to 1100 nm). The inset in Fig. 5.10 shows that the improvement in EQE, due to attached FEP and FEP MCA, decreases with increasing wavelength and is most pronounced in the visible spectrum between 400 nm and 700 nm. Because the solar photon flux is highest in this spectral region, the improvement results in a significant increase in electrical gain. For normal incidence, both the EQE and I-V curves indicate that the introduction of FEP has a bigger impact on the electrical gain of PV modules than the microstructure of FEP MCA. A possible reason for the less pronounced impact of the MCA microstructure are unstructured ridges between the microcavities (Fig. 3.8 c), d)), which possibly increase reflection and decrease the electrical gain. By fine tuning the fabrication parameters, these ridges can be eliminated, minimizing reflectivity, and, thus, maximizing the electrical gain.

As mentioned before, PV modules are installed at various different tilt angles, depending on the installation site and the surrounding landscape. Therefore, sunlight usually does not hit modules under normal incidence but rather at various different AOI. To investigate the effect of varying tilt angles on the reflectivity of PV modules, the total reflection of a bare solar cell, a solar cell coated with unstructured FEP and a cell coated with FEP MCA illuminated at 8° and 80° was measured as shown in Fig. 5.11. The reflection spectra at 8° verify the result of the EQE and I-V curve measurements under normal incidence. Introduction of unstructured FEP and FEP MCA results in a broadband reduction in reflectivity compared to the uncoated device with the smallest difference at a wavelength of λ =650 nm, for which the anti-reflective coating of the solar cell has been optimized. For decreasing and increasing wavelengths however, FEP and FEP MCA significantly reduce the reflectivity with the highest difference between coated and uncoated device being observed in the lower wavelength range. At 8° tilt, almost no difference in reflectivity can be observed between unstructured FEP and FEP MCA (Fig. 5.11 a). However, with increasing AOI the reflectivity of the bare solar cell and the cell coated with unstructured FEP greatly increases ($\approx$ 20% and $\approx$ 35% at 650 nm respectively), while the reflectivity of the FEP MCA coated cell hardly changes ($\approx$ 13% at 650 nm). Thus, reflectivity of the PV device coated with unstructured FEP at 80° is three times as high as reflectivity of the FEP MCA coated device and even higher than on the uncoated device. On the other hand, FEP MCA shows anti-reflective behavior over the complete spectrum relevant for electrical conversion. Because unstructured FEP drastically increases the reflectivity, the anti-reflective behavior can solely be attributed to the embossed microstructure. Hence, for real-world applications, where sunlight often incides on PV modules at an AOI greater than 8°, FEP MCA shows superior optical properties compared to both alternatives, while additionally displaying self-cleaning properties.

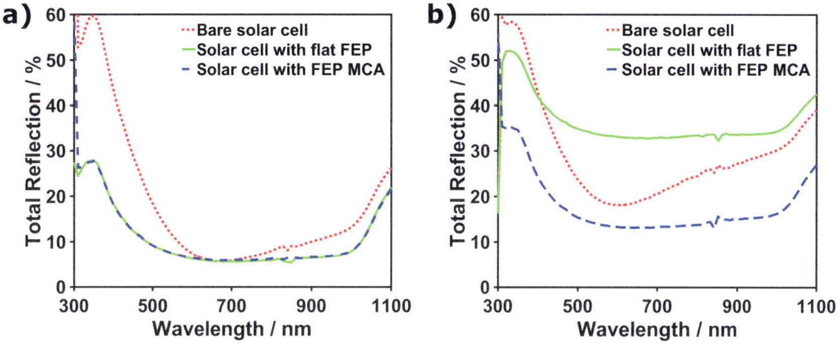

Figure 5.11: Reflectivity of a FEP MCA coated PV device. Total reflection as a function of wavelength for a bare glass-encapsulated solar cell, a cell coated with unstructured FEP and a cell coated with FEP MCA illuminated under a) 8° and b) 80°. With increasing AOI reflectivity of a FEP MCA coated PV module is significantly lower than reflectivity of a bare and unstructured FEP coated PV device, indicating anti-reflective properties of FEP MCA.

5.3 Conclusion

Inspired by the dense layer of transparent hairs on water plants *Salvinia* and *Pistia stratiotes*, which result in superhydrophobicity without impairing optical transmission, the optical and self-cleaning properties of artifical thin nanofur made from optical grade PC were investigated. Superhydrophobicity of thin nanofur results in self-cleaning properties against a variety of contaminants necessary for reducing soiling of PV devices and various other technical applications. When used as a coating, thin nanofur was shown to exhibit reflection values as low as 4%. Furthermore, the high mechanical flexibility makes thin nanofur applicable as a superhydrophobic coating not only on rigid but also on curved and flexible surfaces such as ultrathin flexible solar cells. As a self-standing film, thin nanofur was shown to exhibit transmission values above 85%, with 97% of transmitted light being scattered in a forward direction. Attaching thin nanofur to a monochromatic OLED was shown to increase the out-coupling efficiency by more than 10% compared to an uncoated and unstructured PC coated device. Additionally, the low

front reflectance and high back reflection of thin nanofur increases the short-circuit current density of mc-Si solar cells by 5.8% compared to a bare device.
To overcome limitations in applicabilty of thin nanofur due to low mechanical, chemical and UV stability, while keeping its excellent optical properties, its optically relevant surface topography was transferred to a fluorinated polymer. To this effect, the optical properties of thin nanofur were studied using ray tracing simulations and a MCA topography was shown to significantly reduce reflectivity for all AOI compared to a combination of μ-cavities and nano- and microhairs. Hence, the MCA topography was transferred into FEP and characterized for its applicability on optoelectronic devices. Self-cleaning properties of FEP MCA are demonstrated for several contaminant species and a wide range of contaminant sizes (30 μm - 900 μm) and tilt angles (10° - 45°), thus, including parameters encountered in real-world applications. Furthermore, rain droplets impacting FEP MCA are shown to easily bounce off the surface independent of tilt angle and drop velocity, demonstrating high impact stability of the surface and its wetting properties. Moreover, a 4.6% relative gain in short-circuit current of glass encapsulated mc-Si solar cells was achieved by optically coupling FEP MCA to the PV device. Anti-reflective properties of the embossed MCA at high AOI promises this efficiency enhancement to become even more prominent for PV modules installed at angles differing from normal incidence, as is the case for practical applications. Based on these results, further angle resolved EQE measurements of FEP MCA are of great interest for optoelectronic applications and will give a more detailed insight into the effect of FEP MCA on PV modules. Such investigations will show superiority of the embossed films compared to the conventional glass encapsulation. Furthermore, an investigation into foregoing glass encapsulation of PV modules in favor of FEP MCA is a promising approach to increase efficiency, while simultaneously reducing manufacturing costs. Direct FEP MCA encapsulation would not only reduce manufacturing costs by reducing a two-step (glass encapsulation + coating the PV device) to a one-step fabrication process, but would additionally eliminate one solid-solid interface, which could reduce reflection losses and increase electrical gain. Moreover, self-cleaning properties of the

films combined with their high UV, mechanical and chemical resistance would make an exchange of the FEP MCA coating unnecessary, resulting in long life cycles of PV modules with minimal maintenance.

A shortened version of the optical characterization and application of thin nanofur (Section 5.1) was published in "Bioinspired Superhydrophobic Highly Transmissive Films for Optical Applications", Felix Vüllers, Guillaume Gomard, Jan B. Preinfalk, Efthymios Klampaftis, Matthias Worgull, Bryce S. Richards, Hendrik Hölscher and Maryna N. Kavalenka, Small, 2016, 12, 6144-6152.

A shortened version of the optical characterization and application of FEP MCA (Section 5.2) was published in "Self-Cleaning Microcavity Array for Photovoltaic Modules", Felix Vüllers, Benjamin Fritz, Aiman Roslizar, Andreas Striegel, Markus Guttmann, Bryce S. Richards, Hendrik Hölscher, Guillaume Gomard, Efthymios Klampaftis and Maryna N. Kavalenka, ACS Applied Materials & Interfaces, 2018, 10, 2929-2936.

6 Summary and Outlook

Superhydrophobic surfaces, both natural and artificial, exhibit a great variety of functionalities, which can be exploited for a multitude of energy harvesting and energy conserving applications. In order to harness the enormous potential of superhydrophobic surfaces classical micro- and nanostructuring techniques were used to fabricate superhydrophobic surfaces, which were then examined in the context of fluidic drag reduction and optoelectronic applications. The scientific objectives of this work could thereby be fulfilled and the main findings are summarized in the following.

Development of a scalable and cost efficient fabrication technique for transparent superhydrophobic thin nanofur.

Based on "thick nanofur" [50, 51], a two-step fabrication procedure combining hot embossing and hot pulling was developed to fabricate a thin nanofur film with a minimal base layer thickness of 30 µm, suppporting a dense layer of high aspect ratio micro- and nanohairs. Bonding biocompatible optical grade PC to a sacrificial polymer layer before structuring thin nanofur, minimizes its backside roughness, and, thus, reduces reflection and absorption losses, essential for high transmission and its use as a functional coating on optoelectronic devices. The resulting superhydrophobic thin nanofur exhibits static water contact angles above 150°, roll-off angles below 6° and retains an air layer when submerged underwater. Furthermore, by varying the fabrication parameters or perforating thin nanofur by laser drilling or manual perforation, thin nanofur can be functionalized and its air retaining capabilities improved, without changing its wetting properties. Exposition to argon plasma is shown to temporarily reverse the wetting properties

of thin nanofur to underwater superoleophobic. By adjusting the storing conditions of the treated films, the hydrophobic recovery back to superhydrophobicity can be delayed or even halted, in case the treated films are stored in water.
To increase mechanical, UV and chemical resistance, and decrease reflection losses, which is essential for application on optoelectronic devices, a fabrication route was developed to imprint the underlying microcavity array (MCA) of thin nanofur into the PV-relevant polymer fluorinated ethylene propylene (FEP). By electroplating thin nanofur and then hot embossing the fabricated Ni-shim into FEP, a superhydrophobic thin FEP MCA film with water contact angles above 150° and water sliding angles below 5° was fabricated.
In conclusion, developed thin nanofur and nanofur-based MCA are applicable for drag reduction and optoelectronics. However, to be suitable for industrial application, some open issues have to be tackled first. Optimizing the FEP MCA fabrication parameters can increase the AR of the µ-cavities, eliminate unstructured ridges between the cavities, and, thus, reduce reflection losses. Furthermore, a reliable high throughput fabrication procedure of both FEP MCA and thin nanofur has to be established. A transfer to a roll-to-roll replication procedure has great potential, as the fabrication procedure of both films is based on hot embossing and hot pulling techniques, and fabrication of the respective shims is inexpensive and low in complexity. The greatest challenge for large scale fabrication of thin nanofur is cleaning of the sandblasted steel drum used as a mold insert during hot pulling. A possible solution for removing polymer from the drum could be *in-situ* chemical or plasma cleaning of the utilized steel drum. Moreover, current projects at the Institute of Microstructure Technology target developement of seamless sleeves for roll-to-roll fabrication. Such shims will overcome limitations in shim size which limit FEP MCA fabrication to conventional hot embossing, rather than roll-to-roll fabrication. An upscaling of the fabrication procedures of thin nanofur and FEP MCA would greatly increase throughput and enable coating of large areas, such as entire ships or PV arrays with functionalized superhydrophobic surfaces, thus, pushing towards solutions for such problems as

increasing fuel consumption in international shipping and the increasing need for renewable energy.

Equipment of thin nanofur with a highly stable underwater retained air layer.

Analogous to its natural prototypes, the dense layer of high aspect ratio micro- and nanohairs enables thin nanofur to retain an air layer when submerged under water. By characterizing the wetting transition on superhydrophobic nanofur, it was shown that its superhydrophobic topography with moderate surface energy retains an air layer which is more stable against hydrostatic pressure than the same topography with either low or high surface energy. Additionally, an easy to use measurement technique based on total internal reflection at the air-water interface was introduced to show that the air covered area on thin nanofur decreases exponentially with increasing hydrostatic pressure. To overcome the pressure limitation of the retained air layer, nanofur was perforated and its underwater retained air layer was pressurized. Air layer pressurization counteracts the hydrostatic pressure acting on the air-water interface, and results in retained air layer stable up to 7 bar hydrostatic pressure and pressure fluctuations of 3 bar.

Moreover, by combining the total internal reflection measurement technique with contact line motion observations, it was possible to identify six topographical features necessary for high dynamic plastron stability and high contamination resistance of superhydrophobic surfaces. These features are: (i) hierarchical topography; (ii) high density of surface structures; (iii) minimal water-solid contact area; (iv) nanometer-scale features; (v) cavities or cavity-like structures and (vi) re-entrant structures. Moreover, these topographical features were correlated to the dynamic wetting behavior of superhydrophobic surfaces and a discrepancy between stability enhancing and drag reducing features was shown.

Both developed measurement techniques are easy to use and easily adjustable to a variety of different measurement environments, thus, providing a good foundation for characterization of the retained air layer stability, both in static and dynamic conditions, often neglected in other existing research. A further investigation of stability enhancing topographical features (dynamic and static) and an in-depth investigation into the drag reducing capabilities of perforated thin nanofur is of great interest. The high static stability of the retained air layer on pressure supported perforated thin nanofur has the potential to reduce frictional drag under extreme conditions often encountered in harsh environments. Measurements of the drag-reducing capabilities of thin nanofur and perforated thin nanofur after integration into model ship surfaces or microfluidic devices could be the next step.

Application of self-cleaning thin nanofur and nanofur-based films as efficiency enhancing coatings on optoelectronic devices.

The superhydrophobicity of thin nanofur results in a self-cleaning behavior against a variety of contaminants with a broad range of particle sizes. Self-cleaning behavior is beneficial for reducing soiling of optoelectronic devices. An in-depth optical characterization of bioinspired thin nanofur shows that it exhibits reflection values of less than 4% when used as a coating, while more than 86% of inciding light is scattered in a forward direction, when used as a self-standing film. Such excellent optical properties are beneficial for light managing applications, and are shown to increase the out-coupling efficiency of OLEDs by up to 14%, and also enhance the efficiency of a bare mc-Si solar cells by almost 6%.

Optical coupling of nanofur-based FEP MCA to a glass encapsulated mc-Si solar cell leads to a 4.6% increase in electrical gain under normal incidence, compared to a bare glass encapsulated cell. Furthermore, numerical simulations show that the bare MCA topography exhibits lower reflection values than thin nanofur for all AOI. This anti-reflective behavior of FEP MCA enhances drastically with increasing AOI, leading to an up to 50% reduced reflectivity compared to unstructured FEP at an AOI of 80°. Together with the self-cleaning properties shown for a broad range of contaminant species, sizes and tilt angles, as well as the surface

stability against impacting water droplets, the anti-reflective behavior makes FEP MCA an effective coating for light management.
A further investigation into the benefits of FEP MCA for PV efficiency at different tilt angles, as well as a field testing of self-cleaning properties and UV- and mechanical stability would be of great interest. Moreover, foregoing the glass encapsulation of PV modules in favor of a direct coating with FEP MCA might further reduce reflection losses and increase the cost efficiency of coated PV modules. Finally, the high mechanical flexibility and translucent properties of both thin nanofur and FEP MCA have great potential for implementation into ultrathin flexible solar cells, where forward scattering increases the optical path in the active medium, thus, further boosting the absorption probability and electrical gain.

Appendix

Semi-Automatic Calculation of Dynamic Contact Angle

```
% Filename: ContactLineMotion.m
% Author: Felix Vuellers
% Date: 01.03.2018
% Karlsruhe Institute of Technology (KIT)
% Matlab Version: MatlabR2015b

function [angle_deg, cleansed_angle] ...
    = ContactLineMotion()


% Path of Contact Line Motion Images
direct='D:\Users\hb3224\Documents\Dip01_50mmmin...
    \sliced\crop\up\images\';
fpathdata=direct;


% Getting all contact line motion files with
% specified extension;
getAllTiff=getAllFiles(fpathdata,'*.tiff',true);
if isempty(getAllTiff)
    getAllTiff=getAllFiles(fpathdata,'*.tif',true);
end
getAllTiff=char(getAllTiff);


% Define number of points not taken into
```

```
% consideration at beginning due to artifacts
point_omit=2;

% Definition of Threshold intensity, defining interface
intensity_threshold=70;

% Definition of number of points used to calculate
% linear regression for tangent
tangent_points=8;

% Define deviation that is allowed from mean value in
% contact angle without being labeled as program not
%working and moved for manual reevaluation
allowed_deviation=10;

% Define after how many frames the coordinates of the
% substrate should be reset by clicking
click_reset=5;

% Create folders to save pictures fabricated in measurement,
% if not existent
if(~exist(direct,'dir'));
    mkdir(direct);
end

help=strcat(direct,'Overlap');
if (~exist(help,'dir'))
    mkdir(help);
    mkdir(help,'Tangent');
end
```

```
for i=1:size(getAllTiff,1)
    % Resave name of currently used picture in order
    % to make handling easier
    fname=getAllTiff(i,:);

    % Break up name into filename and rest of name
    % --> only filename necessary
    [~,filename,~] = fileparts(fname);

    % Create strings to later save data and pictures
    % in appropriate folders
    AWI_tangent = ...
        strcat(direct,'Overlap\Tangent\',filename,'.tif');
    ContactAngleFile=strcat(direct,'_ContactAngle.mat');

    % loads the picture to extract data from it
    load_crosssection = imread(fname);

    % To find the location of the pixel reaching
    % threshold intensity;
    [rows,cols] = ...
        find(load_crosssection<intensity_threshold);

    % location of pixel using binary image
    Y = [rows,cols];

    % information about the loaded image
    % width = size x-direction (number of pictures in
    % original stack) and
    % heigth = size y-direction (length of edge
    % of one picture in original stack)
```

```
info = imfinfo (fname);
w = info(1).Width;
h = info(1).Height;

%Creates an empty 3-D array
intensity_profile = zeros(h,w,3,'uint8');
intensity_profile(:,:,:)=255;
Trace = zeros(h,w,3,'uint8');
S=zeros(h,1);
N=zeros(h,1);

% Visualization of the intensity profile in RGB
% Sets every pixel in Y (every pixel exceeding the
% threshold to some arbitrary color in RGB values.
% Only first value of intensity profile will be used
% in later script to determine further computations)
for j=1:size(Y,1)
    intensity_profile(Y(j,1),Y(j,2),1)=200;
    intensity_profile(Y(j,1),Y(j,2),2)=75;
    intensity_profile(Y(j,1),Y(j,2),3)=50;
end

for k=1:h
    for l=1:w
        if intensity_profile(k,l,1)==255 && S(k)==0
           % Goes through the rows of the image from
           % left to right; for every pixel that
           % doesn't exceed the threshold with no
           % pixel in that row before it exceeding
           % the threshold it increases the
           % respective N by 1 (not above 255); i.e.
```

```
            % the first row contains 5 pixels with third
            % and fifth exceeding the threshold
            % -->N(1)=2 (increased at first and second
            % iteration, doesn't increase for 4th pixel,
            % since the third increases S-->S(1)
            % unequal 0); i.e. the first row contains
            % 5 pixels with first and fifth exceeding
            % the threshold -->N(1)=0 (first pixel
            % increases S --> S(1) unequal 0 for the
            % rest of the pixels --> N doesn't increase;
             N(k)=N(k)+1;
        elseif intensity_profile(k,l,1)==200 && S(k)==0
             % Similar to N(k) determination.
             % It will become one for the first pixel
             % that reaches the threshold. Afterwards
             % no more changes are made to S(k) or N(k)
             %in that specific row
             S(k)=S(k)+1;
        end
    end
    if S(k)~=0
        % colors the first pixel reaching the threshold
        % N(k) is number of pixels starting from left,
        % that are not higher than threshold and S(k)/2
        % is the first pixel reaching threshold due to
        % upper definition i.e. first row has 20 pixels
        % with first 7 below threshold then 6 higher
        % and then 4 below --> N(1)=7 S(1)=6 -->
        % N(k)+S(k)/2=13 = middle of bright area.
        % This calculation however only works,
        % if the area higher than threshold is
```

```
        % continuous, otherwise S(k) does not
        % represent the thickness of the interface
        % --> will not work for more than one interface
        Trace(k,N(k)+round(S(k)/2),1)=0;
        Trace(k,N(k)+round(S(k)/2),2)=250;
        Trace(k,N(k)+round(S(k)/2),3)=0;
    end
end

% Trace is the rgb matrix for tracing the outline of
% the interface and I is the two dimensional matrix
% in greyscale
I=rgb2gray(Trace);
[rows,cols] = find(I);
Z = [rows,cols];
D = sortrows(Z);

% Highlighting the extracted air-water interface in
% the image
overlap = repmat(load_crosssection,[1,1,3]);
for r=(find(cols>1,1):(length(rows)-1))
    overlap(rows(r),cols(r)-1,1)=0;
    overlap(rows(r),cols(r)-1,3)=0;
    overlap(rows(r),cols(r)-1,2)=250;
    overlap(rows(r),cols(r),1)=0;
    overlap(rows(r),cols(r),3)=0;
    overlap(rows(r),cols(r),2)=250;
    overlap(rows(r),cols(r)+1,1)=0;
    overlap(rows(r),cols(r)+1,3)=0;
    overlap(rows(r),cols(r)+1,2)=250;
end
```

```
% Define solid-liquid-vapor-triple point manually;
% Find closest point to contact line by clicking
% next to it plotting overlap with fitted polynomial
% and tangent
h1=figure('Name',sprintf('Click number %s %s', ...
    num2str(i), filename(1,:)),'NumberTitle','Off',...
    'Position', [162 55 867 612],'Visible','on');
imshow(overlap);
set(h1,'Position',[162 55 867 612]);
hold on
if (i==1 || mod(i,click_reset)==1 || click_reset==1)
    [x,y]=ginput(1);
end

closest_point_x=D(find(abs(D(:,1)-y)<4),:);
closest_point=closest_point_x(...
    find(abs(closest_point_x(:,2)-x)<4),:);

% In case contact line moves too far away
% from first click, new position will be defined
while isempty(closest_point)
    [x,y]=ginput(1);
    closest_point_x=D(find(abs(D(:,1)-y)<4),:);
    closest_point=...
        closest_point_x(find(...
            abs(closest_point_x(:,2)-x)<4),:);
end
% Find first index of the substrate line
first_index_substrate=find(...
    closest_point(:,2)==closest_point(end,2),1);
```

```
D_substrate_index=...
    closest_point(first_index_substrate,1);

help_x=[D_substrate_index:(-1):...
    D_substrate_index()-tangent_points];
help_y=transpose(D(help_x,2));
p=polyfit(help_x,help_y,1);
slope_red=p(1);
x_tangent=[0:0.1:200];
tangent_red=slope_red*(x_tangent-(D_substrate_index))...
    +polyval(p,D_substrate_index);

% calculating ContactAngle
v_tangent=[1 slope_red];
% Polynomial is calculated for a normal orientation of
% the axis. Therefore the angle is in respect to a
% horizontal line instead of a vertical line as seen
% in the picture from overlap (overlap has a 90Â°
% tilted view)
v_vert=[1 0];
angle(i)=atan2(abs(det([transpose(v_vert),...
    transpose(v_tangent)])),dot(transpose(v_vert),...
    transpose(v_tangent)));
angle_deg(i)=rad2deg(angle(i));
angle_deg(i)=180-angle_deg(i);

% As the tangent is also calculated in the rotated
% system, the change of variables has to be done
% here as well
plot(tangent_red,x_tangent,'cyan');
```

```
    % increase tangent width for better visibility
    help=get(gca,'children');
    set(help(1),'LineWidth',2.5);
    saveas(h1,AWI_tangent);
    close(h1);
end

angle_deg=transpose(angle_deg);

% Move files that differ too much from mean value
% to different folder to be reevaluated
mean_angle_deg=mean(angle_deg);index_trouble=...
    find(abs(angle_deg-mean_angle_deg)>allowed_deviation);

moved_files=getAllTiff(index_trouble,:);

if ~isempty(moved_files)
    % create new folder and move deviating
    % files into that folder
    folder_help=strcat(direct,'reevaluate');
    mkdir(folder_help);

    for ll=1:size(moved_files,1)
        copyfile(moved_files(ll,:),folder_help);
    end

    % Save names of files to reevaluate and
    % their indexes to later see which contact angle
    % has been replaced
    name_length=size(moved_files,2)
    fileID = fopen(strcat(direct,'_reevaluate.txt'),'w');
```

```
    fprintf(fileID,'%20s \r\n','Files reevaluated');
    for r=1:size(moved_files,1)
        fprintf(fileID,sprintf(...
            '%s%i%s %s %s%i%s \r\n','%',name_length,...
            's', '  ', '%',3,'i'),...
            moved_files(r,:),index_trouble(r));
    end
    fclose(fileID);

    cleansed_angle_deg=angle_deg(find(abs(...
        angle_deg-mean_angle_deg)<=allowed_deviation));
    % save contact angles
    save(ContactAngleFile,'angle_deg',...
        'cleansed_angle_deg','index_trouble');
else
    % save contact angles
    save(ContactAngleFile,'angle_deg');
end

end
```

Bibliography

[1] Barthlott, W., and Neinhuis, C. Purity of the sacred lotus, or escape from contamination in biological surfaces, *Planta*, **1997**, *202*, 1–8.

[2] Barthlott, W., Schimmel, T., Wiersch, S., Koch, K., Brede, M., Barczewski, M., Walheim, S., Weis, A., Kaltenmaier, A., Leder, A., and Bohn, H. F. The Salvinia Paradox: Superhydrophobic Surfaces with Hydrophilic Pins for Air Retention Under Water, *Adv. Mater.*, **2010**, *22*, 2325–2328.

[3] Neinhuis, C., and Barthlott, W. Characterization and Distribution of Water-repellent, Self-cleaning Plant Surfaces, *Ann. Bot.*, **1997**, *79*, 667–677.

[4] Mayser, M. J., and Barthlott, W. Layers of Air in the Water beneath the Floating Fern Salvinia are Exposed to Fluctuations in Pressure, *Integr. Comp. Biol.*, **2014**, *54*, 1001–1007.

[5] Cerman, Z., Striffler, B. F., and Barthlott, W. *Functional Surfaces in Biology*; Springer Science + Business Media, **2009**; pp 97–111.

[6] Genzer, J., and Efimenko, K. Recent developments in superhydrophobic surfaces and their relevance to marine fouling: a review, *Biofouling*, **2006**, *22*, 339–360.

[7] Kim, J.-G., Choi, H. J., Park, K.-C., Cohen, R. E., McKinley, G. H., and Barbastathis, G. Multifunctional Inverted Nanocone Arrays for Non-Wetting, Self-Cleaning Transparent Surface with High Mechanical Robustness, *Small*, **2014**, *10*, 2487–2494.

[8] Min, W.-L., Jiang, B., and Jiang, P. Bioinspired Self-Cleaning Antireflection Coatings, *Adv. Mater.*, **2008**, *20*, 3914–3918.

[9] Koch, K., Bhushan, B., Jung, Y. C., and Barthlott, W. Fabrication of artificial Lotus leaves and significance of hierarchical structure for superhydrophobicity and low adhesion, *Soft Matter*, **2009**, *5*, 1386–1393.

[10] Ditsche-Kuru, P., Schneider, E. S., Melskotte, J.-E., Brede, M., Leder, A., and Barthlott, W. Superhydrophobic surfaces of the water bug Notonecta glauca: a model for friction reduction and air retention, *Beilstein J. Nanotechnol.*, **2011**, *2*, 137–144.

[11] Barthlott, W., Mail, M., and Neinhuis, C. Superhydrophobic hierarchically structured surfaces in biology: evolution, structural principles and biomimetic applications, *Philos. Trans. R. Soc., A*, **2016**, *374*, 0191–0231.

[12] Papageorgiou, D. P., Tsougeni, K., Tserepi, A., and Gogolides, E. Superhydrophobic, hierarchical, plasma-nanotextured polymeric microchannels sustaining high-pressure flows, *Microfluid. Nanofluid.*, **2013**, *14*, 247–255.

[13] Choi, C.-H., and Kim, C.-J. Large Slip of Aqueous Liquid Flow over a Nanoengineered Superhydrophobic Surface, *Phys. Rev. Lett.*, **2006**, *96*, 066001–066004.

[14] Bixler, G. D., and Bhushan, B. Fluid Drag Reduction with Shark-Skin Riblet Inspired Microstructured Surfaces, *Adv. Funct. Mater.*, **2013**, *23*, 4507–4528.

[15] Srinivasan, S., Kleingartner, J. A., Gilbert, J. B., Cohen, R. E., Milne, A. J. B., and McKinley, G. H. Sustainable Drag Reduction in Turbulent Taylor-Couette Flows by Depositing Sprayable Superhydrophobic Surfaces, *Phys. Rev. Lett.*, **2015**, *114*, 014501–014505.

[16] Bhushan, B., and Jung, Y. C. Natural and biomimetic artificial surfaces for superhydrophobicity, self-cleaning, low adhesion, and drag reduction, *Prog. Mater Sci.*, **2011**, *56*, 1–108.

[17] Zhang, P., and Lv, F. A review of the recent advances in superhydrophobic surfaces and the emerging energy-related applications, *Energy*, **2015**, *82*, 1068–1087.

[18] Yao, X., Song, Y., and Jiang, L. Applications of Bio-Inspired Special Wettable Surfaces, *Adv. Mater.*, **2011**, *23*, 719–734.

[19] Gogolides, E., Ellinas, K., and Tserepi, A. Hierarchical micro and nano structured, hydrophilic, superhydrophobic and superoleophobic surfaces incorporated in microfluidics, microarrays and lab on chip microsystems, *Microelectron. Eng.*, **2015**, *132*, 135–155.

[20] Sackmann, E. K., Fulton, A. L., and Beebe, D. J. The present and future role of microfluidics in biomedical research, *Nature*, **2014**, *507*, 181–189.

[21] Eyring, V., Köhler, H. W., van Aardenne, J., and Lauer, A. Emissions from international shipping: 1. The last 50 years, *J. Geophys. Res.*, **2005**, *110*, 305–316.

[22] Buhaug, Ø., Corbett, J., Endresen, O., Eyring, V., Faber, J., Hanayama, S., Lee, D., Lee, D., Lindstad, H., Mjelde, A., Palsson, C., Wanquing, W., Winebrake, J., and Yoshida, K. Second IMO greenhouse gas study, *International Maritime Organization, London*, **2009**,

[23] Ceccio, S. L. Friction Drag Reduction of External Flows with Bubble and Gas Injection, *Annu. Rev. Fluid Mech.*, **2010**, *42*, 183–203.

[24] Gokcay, S., Insel, M., and Odabasi, A. Revisiting artificial air cavity concept for high speed craft, *Ocean Eng.*, **2004**, *31*, 253–267.

[25] Kawaguchi, Y., Segawa, T., Feng, Z., and Li, P. Experimental study on drag-reducing channel flow with surfactant additives-spatial structure of turbulence investigated by PIV system, *Int. J. Heat Fluid Flow*, **2002**, *23*, 700–709.

[26] Golovin, K. B., Gose, J. W., Perlin, M., Ceccio, S. L., and Tuteja, A. Bioinspired surfaces for turbulent drag reduction, *Philos. Trans. R. Soc., A*, **2016**, *374*, 189–208.

[27] Wang, G., Guo, Z., and Liu, W. Interfacial Effects of Superhydrophobic Plant Surfaces: A Review, *J. Bionic Eng.*, **2014**, *11*, 325–345.

[28] Suhr, S. H., Song, Y. S., Lee, S. J., and Sitti, M. Biologically Inspired Miniature Water Strider Robot., Robotics: Science and Systems, **2005**; pp. 319–326.

[29] Brassard, J.-D., Sarkar, D., and Perron, J. Studies of drag on the nanocomposite superhydrophobic surfaces, *Appl. Surf. Sci.*, **2015**, *324*, 525–531.

[30] Jung, Y. C., and Bhushan, B. Biomimetic structures for fluid drag reduction in laminar and turbulent flows, *J. Phys. Condens. Matter*, **2010**, *22*, 035104–035112.

[31] Lyu, S., Nguyen, D. C., Kim, D., Hwang, W., and Yoon, B. Experimental drag reduction study of super-hydrophobic surface with dual-scale structures, *Appl. Surf. Sci.*, **2013**, *286*, 206–211.

[32] Zhang, H., Yin, L., Li, L., Shi, S., Wang, Y., and Liu, X. Wettability and drag reduction of a superhydrophobic aluminum surface, *RSC Adv.*, **2016**, *6*, 14034–14041.

[33] Ou, J., Perot, B., and Rothstein, J. P. Laminar drag reduction in microchannels using ultrahydrophobic surfaces, *Phys. Fluids*, **2004**, *16*, 4635–4643.

[34] Sarver, T., Al-Qaraghuli, A., and Kazmerski, L. L. A comprehensive review of the impact of dust on the use of solar energy: History, investigations, results, literature, and mitigation approaches, *Renew. Sustain. Energy Rev.*, **2013**, *22*, 698–733.

[35] Mani, M., and Pillai, R. Impact of dust on solar photovoltaic (PV) performance: Research status, challenges and recommendations, *Renew. Sustain. Energy Rev.*, **2010**, *14*, 3124–3131.

[36] Darwish, Z. A., Kazem, H. A., Sopian, K., Al-Goul, M., and Alawadhi, H. Effect of dust pollutant type on photovoltaic performance, *Renew. Sustain. Energy Rev.*, **2015**, *41*, 735–744.

[37] Costa, S. C., Diniz, A. S. A., and Kazmerski, L. L. Dust and soiling issues and impacts relating to solar energy systems: Literature review update for 2012-2015, *Renew. Sustain. Energy Rev.*, **2016**, *63*, 33–61.

[38] Jamil, W. J., Rahman, H. A., Shaari, S., and Salam, Z. Performance degradation of photovoltaic power system: Review on mitigation methods, *Renew. Sustain. Energy Rev.*, **2017**, *67*, 876–891.

[39] Micheli, L., and Muller, M. An investigation of the key parameters for predicting PV soiling losses, *Prog. Photovoltaics Res. Appl.*, **2017**, *25*, 291–307.

[40] Sayyah, A., Horenstein, M. N., and Mazumder, M. K. Energy yield loss caused by dust deposition on photovoltaic panels, *Sol. Energy*, **2014**, *107*, 576–604.

[41] Ragesh, P., Anand Ganesh, V., Nair, S. V., and Nair, A. S. A review on 'self-cleaning and multifunctional materials', *J. Mater. Chem. A*, **2014**, *2*, 14773–14797.

[42] Sun, C.-H., Gonzalez, A., Linn, N. C., Jiang, P., and Jiang, B. Templated biomimetic multifunctional coatings, *Appl. Phys. Lett.*, **2008**, *92*, 051107–051109.

[43] Siddique, R. H., Gomard, G., and Hölscher, H. The role of random nanostructures for the omnidirectional anti-reflection properties of the glasswing butterfly, *Nat. Commun.*, **2015**, *6*, 6909–6916.

[44] Leem, J. W., Kim, S., Lee, S. H., Rogers, J. A., Kim, E., and Yu, J. S. Efficiency Enhancement of Organic Solar Cells Using Hydrophobic Antireflective Inverted Moth-Eye Nanopatterned PDMS Films, *Adv. Energy Mater.*, **2014**, *4*, 1301315–1301322.

[45] Palumbo, F., Di Mundo, R., Cappelluti, D., and d'Agostino, R. SuperHydrophobic and SuperHydrophilic Polycarbonate by Tailoring Chemistry and Nano-texture with Plasma Processing, *Plasma Processes Polym.*, **2011**, *8*, 118–126.

[46] Gao, C.-Y., Meng, G.-X., Li, X., Wu, M., Liu, Y., Li, X.-Y., Zhao, X., Lee, I., and Feng, X. Wettability of dragonfly wings: the structure detection and theoretical modeling, *Surf. Interface Anal.*, **2013**, *45*, 650–655.

[47] Meng, L.-Y., and Park, S.-J. Effect of fluorination of carbon nanotubes on superhydrophobic properties of fluoro-based films, *J. Colloid Interface Sci.*, **2010**, *342*, 559–563.

[48] Yu, S., Guo, Z., and Liu, W. Biomimetic transparent and superhydrophobic coatings: from nature and beyond nature, *Chem. Commun.*, **2015**, *51*, 1775–1794.

[49] Röhrig, M., Schneider, M., Etienne, G., Oulhadj, F., Pfannes, F., Kolew, A., Worgull, M., and Hölscher, H. Hot pulling and embossing of hierarchical nano- and micro-structures, *J. Micromech. Microeng.*, **2013**, *23*, 105014–105020.

[50] Röhrig, M., Mail, M., Schneider, M., Louvin, H., Hopf, A., Schimmel, T., Worgull, M., and Hölscher, H. Nanofur for Biomimetic Applications, *Adv. Mater. Interfaces*, **2014**, *1*, 1300083–1300092.

[51] Röhrig, M. Fabrication and analysis of bio-inspired smart surfaces. Ph.D. thesis, Karlsruhe Institute of Technology (KIT), Karlsruhe, **2014**; Zugl.: Karlsruhe, KIT, Diss., 2013.

[52] Reiss, H. Warum gibt es Benetzung?, *Phys. Unserer Zeit*, **1992**, *23*, 204–212.

[53] Young, T. An Essay on the Cohesion of Fluids, *Philos. Trans. Roy. Soc. London*, **1805**, *95*, 65–87.

[54] Quéré, D. Rough ideas on wetting, *Physica A*, **2002**, *313*, 32–46.

[55] Roach, P., Shirtcliffe, N. J., and Newton, M. I. Progess in superhydrophobic surface development, *Soft Matter*, **2008**, *4*, 224–240.

[56] Nishino, T., Meguro, M., Nakamae, K., Matsushita, M., and Ueda, Y. The Lowest Surface Free Energy Based on -CF3 Alignment, *Langmuir*, **1999**, *15*, 4321–4323.

[57] Quéré, D. Wetting and Roughness, *Annu. Rev. Mater. Res.*, **2008**, *38*, 71–99.

[58] Moulinet, S., and Bartolo, D. Life and death of a fakir droplet: Impalement transitions on superhydrophobic surfaces, *Eur. Phys. J. E: Soft Matter Biol. Phys.*, **2007**, *24*, 251–260.

[59] Gorb, S. N. *Functional Surfaces in Biology: Little structures with big effects*; Springer Science & Business Media, **2009**; Vol. 1.

[60] Wenzel, R. N. Resistance of solid surfaces to wetting by water, *Ind. Eng. Chem.*, **1936**, *28*, 988–994.

[61] Cassie, A. B. D., and Baxter, S. Wettability of porous surfaces, *Trans. Faraday Soc.*, **1944**, *40*, 546–551.

[62] Marmur, A. Wetting on Hydrophobic Rough Surfaces: To Be Heterogeneous or Not To Be?, *Langmuir*, **2003**, *19*, 8343–8348.

[63] Quéré, D. Surface chemistry: Fakir droplets, *Nat. Mater.*, **2002**, *1*, 14–15.

[64] Callies, M., and Quere, D. On water repellency, *Soft Matter*, **2005**, *1*, 55–61.

[65] Webb, H. K., Crawford, R. J., and Ivanova, E. P. Wettability of natural superhydrophobic surfaces, *Adv. Colloid Interface Sci.*, **2014**, *210*, 58–64.

[66] Quéré, D., Azzopardi, M.-J., and Delattre, L. Drops at Rest on a Tilted Plane, *Langmuir*, **1998**, *14*, 2213–2216.

[67] Otten, A., and Herminghaus, S. How Plants Keep Dry: A Physicist's Point of View, *Langmuir*, **2004**, *20*, 2405–2408.

[68] Solga, A., Cerman, Z., Striffler, B. F., Spaeth, M., and Barthlott, W. The dream of staying clean: Lotus and biomimetic surfaces, *Bioinspiration & Biomimetics*, **2007**, *2*, S126–S134.

[69] Gao, X., and Jiang, L. Biophysics: water-repellent legs of water striders, *Nature*, **2004**, *432*, 36–36.

[70] Barthlott, W., Wiersch, S., Čolić, Z., and Koch, K. Classification of trichome types within species of the water fern Salvinia , and ontogeny of the egg-beater trichomes, *Botany*, **2009**, *87*, 830–836.

[71] Koch, K., and Barthlott, W. Superhydrophobic and superhydrophilic plant surfaces: an inspiration for biomimetic materials, *Philos. Trans. R. Soc., A*, **2009**, *367*, 1487–1509.

[72] Vüllers, F., Gomard, G., Preinfalk, J. B., Klampaftis, E., Worgull, M., Richards, B., Hölscher, H., and Kavalenka, M. N. Bioinspired Superhydrophobic Highly Transmissive Films for Optical Applications, *Small*, **2016**, *12*, 6144–6152.

[73] Hemeda, A. A., and Tafreshi, H. V. General Formulations for Predicting Longevity of Submerged Superhydrophobic Surfaces Composed of Pores or Posts, *Langmuir*, **2014**, *30*, 10317–10327.

[74] Zheng, Q.-S., Yu, Y., and Zhao, Z.-H. Effects of Hydraulic Pressure on the Stability and Transition of Wetting Modes of Superhydrophobic Surfaces, *Langmuir*, **2005**, *21*, 12207–12212.

[75] Xue, Y., Chu, S., Lv, P., and Duan, H. Importance of Hierarchical Structures in Wetting Stability on Submersed Superhydrophobic Surfaces, *Langmuir*, **2012**, *28*, 9440–9450.

[76] Bormashenko, E. Progress in understanding wetting transitions on rough surfaces, *Adv. Colloid Interface Sci.*, **2015**, *222*, 92–103, Reinhard Miller, Honorary Issue.

[77] Amabili, M., Giacomello, A., Meloni, S., and Casciola, C. M. Collapse of superhydrophobicity on nanopillared surfaces, *Phys. Rev. Fluids*, **2017**, *2*, 034202–034217.

[78] Kavalenka, M. N., Vüllers, F., Kumberg, J., Zeiger, C., Trouillet, V., Stein, S., Ava, T. T., Li, C., Worgull, M., and Hölscher, H. Adaptable bioinspired special wetting surface for multifunctional oil/water separation, *Sci. Rep.*, **2017**, *7*, 39970–39979.

[79] Domininghaus, H. *Kunststoffe: Eigenschaften und Anwendungen*; Springer-Verlag, **2007**.

[80] Covestro, Datasheet: Makrolon LED2045, www.plastics.covestro.com, (2017), accessed: 20. February 2018.

[81] Kavalenka, M. N., Vüllers, F., Lischker, S., Zeiger, C., Hopf, A., Röhrig, M., Rapp, B. E., Worgull, M., and Hölscher, H. Bioinspired Air-Retaining Nanofur for Drag Reduction, *ACS Appl. Mater. Interfaces*, **2015**, *7*, 10651–10655.

[82] Material Data Center, Polymerstoffdatenbank, www.materialdatacenter.com, (2018), accessed: 01. January 2018.

[83] Kolew, A. *Heißprägen von Verbundfolien für mikrofluidische Anwendungen*; Schriften des Instituts für Mikrostrukturtechnik am Karlsruher Institut für Technologie (KIT).: Institut für Mikrostrukturtechnik; KIT Scientific Publ., **2012**.

[84] Greer, A. I. M., Vasiev, I., Della-Rosa, B., and Gadegaard, N. Fluorinated ethylene-propylene: a complementary alternative to PDMS for nanoimprint stamps, *Nanotechnology*, **2016**, *27*, 155301–155308.

[85] The Chemours Company, Teflon FEP Fluoroplastic Film, www.chemours.com, accessed: 09. January 2018.

[86] Flisom Flexible Solar Modules, `www.flisom.com/wp-content/uploads/2017/08/Datasheet_eFlex_3.1m.pdf`, accessed: 09. January 2018.

[87] Kotz, F., Arnold, K., Bauer, W., Schild, D., Keller, N., Sachsenheimer, K., Nargang, T. M., Richter, C., Helmer, D., and Rapp, B. E. Three-dimensional printing of transparent fused silica glass, *Nature*, **2017**, *544*, 337–342.

[88] Kim, S. H., Lee, K.-D., Kim, J.-Y., Kwon, M.-K., and Park, S.-J. Fabrication of photonic crystal structures on light emitting diodes by nanoimprint lithography, *Nanotechnology*, **2007**, *18*, 055306–055310.

[89] Kouba, J., Kubenz, M., Mai, A., Ropers, G., Eberhardt, W., and Loechel, B. Fabrication of Nanoimprint stamps for photonic crystals, *J. Phys. Conf. Ser.*, **2006**, *34*, 897–903.

[90] Worgull, M. *Hot embossing: theory and technology of microreplication*; William Andrew, **2009**.

[91] Schroers, J., Pham, Q., and Desai, A. Thermoplastic Forming of Bulk Metallic Glass mdash; A Technology for MEMS and Microstructure Fabrication, *J. Microelectromech. Syst.*, **2007**, *16*, 240–247.

[92] Kavalenka, M. N., Hopf, A., Schneider, M., Worgull, M., and Holscher, H. Wood-based microhaired superhydrophobic and underwater superoleophobic surfaces for oil/water separation, *RSC Adv.*, **2014**, *4*, 31079–31083.

[93] Berliner, E. Process of producing records of sound. **1888**.

[94] Heidari, B., Maximov, I., and Montelius, L. Nanoimprint lithography at the 6 in. wafer scale, *Journal of Vacuum Science & Technology B: Microelectronics and Nanometer Structures Processing, Measurement, and Phenomena*, **2000**, *18*, 3557–3560.

[95] Koesdjojo, M. T., Tennico, Y. H., and Remcho, V. T. Fabrication of a Microfluidic System for Capillary Electrophoresis Using a Two-Stage Embossing Technique and Solvent Welding on Poly(methyl methacrylate) with Water as a Sacrificial Layer, *Anal. Chem.*, **2008**, *80*, 2311–2318.

[96] Ulrich, R., Weber, H. P., Chandross, E. A., Tomlinson, W. J., and Franke, E. A. Embossed Optical Waveguides, *Appl. Phys. Lett.*, **1972**, *20*, 213–215.

[97] Seekamp, J., Zankovych, S., Helfer, A. H., Maury, P., Torres, C. M. S., Böttger, G., Liguda, C., Eich, M., Heidari, B., Montelius, L., and Ahopelto, J. Nanoimprinted passive optical devices, *Nanotechnology*, **2002**, *13*, 581–586.

[98] Kricka, L. J., Fortina, P., Panaro, N. J., Wilding, P., Alonso-Amigo, G., and Becker, H. Fabrication of plastic microchips by hot embossing, *Lab Chip*, **2002**, *2*, 1–4.

[99] Ehrenstein, G. W. *Polymer-Werkstoffe: Struktur, Eigenschaften, Anwendung*, 3rd ed.; Carl Hanser: München, **2011**.

[100] Heckele, M., and Schomburg, W. K. Review on micro molding of thermoplastic polymers, *J. Micromech. Microeng.*, **2004**, *14*, R1–R14.

[101] Mappes, T., Worgull, M., Heckele, M., and Mohr, J. Submicron polymer structures with X-ray lithography and hot embossing, *Microsyst. Technol.*, **2008**, *14*, 1721–1725.

[102] Köhler, M. *Nanotechnologie*; Wiley-VCH Verlag GmbH & Co. KGaA, **2005**; pp. 37–99.

[103] Gatzen, H. H., Saile, V., and Leuthold, J. *Micro and nano fabrication: tools and processes*; Springer, **2015**.

[104] Franz, G. *Niederdruckplasmen und Mikrostrukturtechnik*; Springer-Verlag, **2013**.

[105] Law, K.-Y. In *Surface Wetting : Characterization, Contact Angle, and Fundamentals*, 1st ed.; Zhao, H., Ed.; SpringerLink : Bücher; Springer: Cham, **2016**.

[106] Lander, L. M., Siewierski, L. M., Brittain, W. J., and Vogler, E. A. A systematic comparison of contact angle methods, *Langmuir*, **1993**, *9*, 2237–2239.

[107] Krishnan, A., Liu, Y.-H., Cha, P., Woodward, R., Allara, D., and Vogler, E. A. An evaluation of methods for contact angle measurement, *Colloids Surf., B*, **2005**, *43*, 95–98.

[108] Hisler, V., Jendoubi, H., Hairaye, C., Vonna, L., Le Houérou, V., Mermet, F., Nardin, M., and Haidara, H. Tensiometric Characterization of Superhydrophobic Surfaces As Compared to the Sessile and Bouncing Drop Methods, *Langmuir*, **2016**, *32*, 7765–7773.

[109] Smestad, G. P. *Optoelectronics of solar cells*; Spie Press, **2002**; Vol. 115.

[110] `http://dx.doi.org/10.1520/G0173-03R12`, ASTM Standard G173-03 "Tables for Reference Solar Spectral Irradiances: Direct Normal and Hemispherical on 37° Tilted Surface." DOI: 10.1520/g0173-03r12.

[111] Xue, Y., Lv, P., Lin, H., and Duan, H. Underwater superhydrophobicity: stability, design and regulation, and applications, *Appl. Mech. Rev.*, **2016**, *68*, 030803–030841.

[112] Song, W., Veiga, D. D., Custódio, C. A., and Mano, J. a. F. Bioinspired Degradable Substrates with Extreme Wettability Properties, *Adv. Mater.*, **2009**, *21*, 1830–1834.

[113] Kim, T.-H., Ha, S.-H., Jang, N.-S., Kim, J., Kim, J. H., Park, J.-K., Lee, D.-W., Lee, J., Kim, S.-H., and Kim, J.-M. Simple and Cost-Effective Fabrication of Highly Flexible, Transparent Superhydrophobic Films with Hierarchical Surface Design, *ACS Appl. Mater. Interfaces*, **2015**, *7*, 5289–5295.

[114] Ditsche, P., Gorb, E., Mayser, M., Gorb, S., Schimmel, T., and Barthlott, W. Elasticity of the hair cover in air-retaining Salvinia surfaces, *Appl. Phys. A*, **2015**, *121*, 505–511.

[115] Lee, J., Majidi, C., Schubert, B., and Fearing, R. S. Sliding-induced adhesion of stiff polymer microfibre arrays. I. Macroscale behaviour, *Journal of The Royal Society Interface*, **2008**, *5*, 835–844.

[116] Kustandi, T. S., Samper, V. D., Ng, W. S., Chong, A. S., and Gao, H. Fabrication of a gecko-like hierarchical fibril array using a bonded porous alumina template, *J. Micromech. Microeng.*, **2007**, *17*, N75–N81.

[117] Bliedtner, J., Müller, H., and Barz, A. *Lasermaterialbearbeitung: Grundlagen - Verfahren - Anwendungen - Beispiele*; Carl Hanser Verlag GmbH & Company KG, **2013**.

[118] Kumberg, J. Trennung von Öl/Wasser-Gemischen mit porösem, biomimetischem Nanopelz. M.Sc. thesis, Karlsruhe Institute of Technology (KIT), **2015**.

[119] Zeiger, C., Kumberg, J., Vüllers, F., Worgull, M., Hölscher, H., and Kavalenka, M. N. Selective filtration of oil/water mixtures with bioinspired porous membranes, *RSC Adv.*, **2017**, *7*, 32806–32811.

[120] Vüllers, F., Fritz, B., Roslizar, A., Striegel, A., Guttmann, M., Richards, B. S., Hölscher, H., Gomard, G., Klampaftis, E., and Kavalenka, M. N. Self-Cleaning Microcavity Array for Photovoltaic Modules, *ACS Appl. Mater. Interfaces*, **2018**, *10*, 2929–2936.

[121] Leach, R., Giusca, C., Guttmann, M., Jakobs, P.-J., and Rubert, P. Development of low-cost material measures for calibration of the metrological characteristics of areal surface texture instruments, *CIRP Ann.*, **2015**, *64*, 545–548.

[122] Schanz, G., and Bade, K. Microelectroforming of metals, *Microengineering of Metals and Ceramics Part II: Special Replication Techniques, Automation and Properties*, **2005**, 395–420.

[123] Scarratt, L. R. J., Hoatson, B. S., Wood, E. S., Hawkett, B. S., and Neto, C. Durable Superhydrophobic Surfaces via Spontaneous Wrinkling of Teflon AF, *ACS Appl. Mater. Interfaces*, **2016**, *8*, 6743–6750.

[124] Ishino, C., Reyssat, M., Reyssat, E., Okumura, K., and Quéré, D. Wicking within forests of micropillars, *EPL (Europhysics Letters)*, **2007**, *79*, 56005–56009.

[125] Kronast, M. Nanofur Funtionalization for Increased Air Retention Stability under Hydrostatic Pressure Fluctuations. M.Sc. thesis, Karlsruhe Institute of Technology (KIT), **2017**.

[126] Yong, J., Yang, Q., Chen, F., Bian, H., Du, G., Farooq, U., and Hou, X. Reversible Underwater Lossless Oil Droplet Transportation, *Adv. Mater. Interfaces*, **2015**, *2*, 1400388–1400393.

[127] Wu, D., Wu, S.-z., Chen, Q.-D., Zhao, S., Zhang, H., Jiao, J., Piersol, J. A., Wang, J.-N., Sun, H.-B., and Jiang, L. Facile creation of hierarchical PDMS microstructures with extreme underwater superoleophobicity for anti-oil application in microfluidic channels, *Lab Chip*, **2011**, *11*, 3873–3879.

[128] Oliveira, N. M., Reis, R. L., and Mano, J. F. Superhydrophobic Surfaces Engineered Using Diatomaceous Earth, *ACS Appl. Mater. Interfaces*, **2013**, *5*, 4202–4208.

[129] Chen, F., Liu, J., Cui, Y., Huang, S., Song, J., Sun, J., Xu, W., and Liu, X. Stability of plasma treated superhydrophobic surfaces under different ambient conditions, *J. Colloid Interface Sci.*, **2016**, *470*, 221–228.

[130] Yong, J., Chen, F., Yang, Q., Du, G., Shan, C., Bian, H., Farooq, U., and Hou, X. Bioinspired transparent underwater superoleophobic and anti-oil surfaces, *J. Mater. Chem. A*, **2015**, *3*, 9379–9384.

[131] Mortazavi, M., and Nosonovsky, M. A model for diffusion-driven hydrophobic recovery in plasma treated polymers, *Appl. Surf. Sci.*, **2012**, *258*, 6876–6883.

[132] Bormashenko, E., Chaniel, G., and Grynyov, R. Towards understanding hydrophobic recovery of plasma treated polymers: Storing in high polarity liquids suppresses hydrophobic recovery, *Appl. Surf. Sci.*, **2013**, *273*, 549–553.

[133] Vogler, E. A. Structure and reactivity of water at biomaterial surfaces, *Adv. Colloid Interface Sci.*, **1998**, *74*, 69–117.

[134] Balmert, A., Bohn, H. F., Ditsche-Kuru, P., and Barthlott, W. Dry under water: Comparative morphology and functional aspects of air-retaining insect surfaces, *J. Morphol.*, **2011**, *272*, 442–451.

[135] McHale, G., Newton, M. I., and Shirtcliffe, N. J. Immersed superhydrophobic surfaces: Gas exchange, slip and drag reduction properties, *Soft Matter*, **2010**, *6*, 714–719.

[136] Carlborg, C. F., and van der Wijngaart, W. Sustained Superhydrophobic Friction Reduction at High Liquid Pressures and Large Flows, *Langmuir*, **2011**, *27*, 487–493.

[137] Vüllers, F., Germain, Y., Petit, L.-M., Hölscher, H., and Kavalenka, M. N. Pressure-Stable Air-Retaining Nanostructured Surfaces Inspired by Natural Air Plastrons, *Adv. Mater. Interfaces*, **2018**, *5*, 1800125–1800132.

[138] Kong, L., Chen, X., Yang, G., Yu, L., and Zhang, P. Preparation and characterization of slice-like Cu2(OH)3NO3 superhydrophobic structure on copper foil, *Appl. Surf. Sci.*, **2008**, *254*, 7255–7258.

[139] Liu, T. L., and Kim, C.-J. Turning a surface superrepellent even to completely wetting liquids, *Science*, **2014**, *346*, 1096–1100.

[140] Sheng, X., and Zhang, J. Air layer on superhydrophobic surface underwater, *Colloids Surf., A*, **2011**, *377*, 374–378.

[141] Pedersen, O., and Colmer, T. D. Physical gills prevent drowning of many wetland insects, spiders and plants, *J. Exp. Biol.*, **2012**, *215*, 705–709.

[142] Samaha, M. A., Ochanda, F. O., Tafreshi, H. V., Tepper, G. C., and el Hak, M. G. In situ, noninvasive characterization of superhydrophobic coatings, *Rev. Sci. Instrum.*, **2011**, *82*, 045109–045115.

[143] Samaha, M. A., Tafreshi, H. V., and el Hak, M. G. Sustainability of superhydrophobicity under pressure, *Phys. Fluids*, **2012**, *24*, 112103–112111.

[144] Bobji, M. S., Kumar, S. V., Asthana, A., and Govardhan, R. N. Underwater Sustainability of the "Cassie" State of Wetting, *Langmuir*, **2009**, *25*, 12120–12126.

[145] Poetes, R., Holtzmann, K., Franze, K., and Steiner, U. Metastable Underwater Superhydrophobicity, *Phys. Rev. Lett.*, **2010**, *105*, 166104–166107.

[146] Lv, P., Xue, Y., Shi, Y., Lin, H., and Duan, H. Metastable States and Wetting Transition of Submerged Superhydrophobic Structures, *Phys. Rev. Lett.*, **2014**, *112*, 196101–196105.

[147] Søgaard, E., Andersen, N. K., Smistrup, K., Larsen, S. T., Sun, L., and Taboryski, R. Study of Transitions between Wetting States on Microcavity Arrays by Optical Transmission Microscopy, *Langmuir*, **2014**, *30*, 12960–12968.

[148] Dilip, D., Jha, N. K., Govardhan, R. N., and Bobji, M. Controlling air solubility to maintain "Cassie" state for sustained drag reduction, *Colloids Surf., A*, **2014**, *459*, 217–224.

[149] Kim, J.-H., and Rothstein, J. P. Delayed lubricant depletion on liquid-infused randomly rough surfaces, *Exp. Fluids*, **2016**, *57*, 81–89.

[150] Wang, J., Wang, B., and Chen, D. Underwater drag reduction by gas, *Friction*, **2014**, *2*, 295–309.

[151] Emami, B., Hemeda, A. A., Amrei, M. M., Luzar, A., el Hak, M. G., and Tafreshi, H. V. Predicting longevity of submerged superhydrophobic surfaces with parallel grooves, *Phys. Fluids*, **2013**, *25*, 062108–062124.

[152] Lee, C., Choi, C.-H., and Kim, C.-J. Superhydrophobic drag reduction in laminar flows: a critical review, *Exp. Fluids*, **2016**, *57*, 176–195.

[153] Checco, A., Ocko, B. M., Rahman, A., Black, C. T., Tasinkevych, M., Giacomello, A., and Dietrich, S. Collapse and Reversibility of the Superhydrophobic State on Nanotextured Surfaces, *Phys. Rev. Lett.*, **2014**, *112*, 216101–216105.

[154] Voinov, O. V. Hydrodynamics of wetting, *Fluid Dyn.*, **1976**, *11*, 714–721.

[155] Cox, R. G. The dynamics of the spreading of liquids on a solid surface. Part 1. Viscous flow, *J. Fluid Mech.*, **1986**, *168*, 169–194.

[156] Kim, J.-H., Kavehpour, H. P., and Rothstein, J. P. Dynamic contact angle measurements on superhydrophobic surfaces, *Phys. Fluids*, **2015**, *27*, 032107.

[157] Lhermerout, R., Perrin, H., Rolley, E., Andreotti, B., and Davitt, K. A moving contact line as a rheometer for nanometric interfacial layers, *Nat. Commun.*, **2016**, *7*, 12545–12550.

[158] Lee, T., Charrault, E., and Neto, C. Interfacial slip on rough, patterned and soft surfaces: A review of experiments and simulations, *Adv. Colloid Interface Sci.*, **2014**, *210*, 21–38.

[159] Bocquet, L., and Charlaix, E. Nanofluidics, from bulk to interfaces, *Chem. Soc. Rev.*, **2010**, *39*, 1073–1095.

[160] Bixler, G. D., and Bhushan, B. Biofouling: lessons from nature, *Philos. Trans. R. Soc., A*, **2012**, *370*, 2381–2417.

[161] Callow, M. E., Jennings, A. R., Brennan, A. B., Seegert, C. E., Gibson, A., Wilson, L., Feinberg, A., Baney, R., and Callow, J. A. Microtopographic Cues for Settlement of Zoospores of the Green Fouling Alga Enteromorpha, *Biofouling*, **2002**, *18*, 229–236.

[162] Zhang, X., Wang, L., and Levanen, E. Superhydrophobic surfaces for the reduction of bacterial adhesion, *RSC Adv.*, **2013**, *3*, 12003–12020.

[163] Privett, B. J., Youn, J., Hong, S. A., Lee, J., Han, J., Shin, J. H., and Schoenfisch, M. H. Antibacterial Fluorinated Silica Colloid Superhydrophobic Surfaces, *Langmuir*, **2011**, *27*, 9597–9601.

[164] Simpson, J. T., Hunter, S. R., and Aytug, T. Superhydrophobic materials and coatings: a review, *Rep. Prog. Phys.*, **2015**, *78*, 086501–086514.

[165] Anselme, K., Davidson, P., Popa, A., Giazzon, M., Liley, M., and Ploux, L. The interaction of cells and bacteria with surfaces structured at the nanometre scale, *Acta Biomater.*, **2010**, *6*, 3824–3846.

[166] Epstein, A. K., Hong, D., Kim, P., and Aizenberg, J. Biofilm attachment reduction on bioinspired, dynamic, micro-wrinkling surfaces, *New J. Phys.*, **2013**, *15*, 095018–095030.

[167] Koch, K., Bhushan, B., and Barthlott, W. Multifunctional surface structures of plants: An inspiration for biomimetics, *Prog. Mater Sci.*, **2009**, *54*, 137–178.

[168] Fürstner, R., , Barthlott, W., Neinhuis, C., and Walzel, P. Wetting and Self-Cleaning Properties of Artificial Superhydrophobic Surfaces, *Langmuir*, **2005**, *21*, 956–961.

[169] Bhushan, B. *Biomimetics - Bioinspired Hierarchical-Structured Surfaces for Green Science and Technology*, 2nd ed.; Springer International Publishing, **2016**.

[170] Lai, Y., Huang, J., Cui, Z., Ge, M., Zhang, K.-Q., Chen, Z., and Chi, L. Recent Advances in TiO2-Based Nanostructured Surfaces with Controllable Wettability and Adhesion, *Small*, **2016**, *12*, 2203–2224.

[171] Hecht, E. *Optik*, 5th ed.; Oldenbourg: München, **2009**.

[172] Smith, R. C., and Baker, K. S. Optical properties of the clearest natural waters (200–800 nm), *Appl. Opt.*, **1981**, *20*, 177–184.

[173] Hünig, R., Mertens, A., Stephan, M., Schulz, A., Richter, B., Hetterich, M., Powalla, M., Lemmer, U., Colsmann, A., and Gomard, G. Flower Power: Exploiting Plants' Epidermal Structures for Enhanced Light Harvesting in Thin-Film Solar Cells, *Adv. Opt. Mater.*, **2016**, 1487–1493.

[174] Giannini, A. In *Sustainable Indoor Lighting*; Sansoni, P., Mercatelli, L., and Farini, A., Eds.; Springer London: London, **2015**; Chapter Radiometric and Photometric Quantities and Laws, pp 1–22.

[175] Möller, S., and Forrest, S. R. Improved light out-coupling in organic light emitting diodes employing ordered microlens arrays, *J. Appl. Phys.*, **2002**, *91*, 3324–3327.

[176] Saxena, K., Jain, V., and Mehta, D. S. A review on the light extraction techniques in organic electroluminescent devices, *Opt. Mater.*, **2009**, *32*, 221–233.

[177] Wrzesniewski, E., Eom, S.-H., Cao, W., Hammond, W. T., Lee, S., Douglas, E. P., and Xue, J. Enhancing Light Extraction in Top-Emitting Organic Light-Emitting Devices Using Molded Transparent Polymer Microlens Arrays, *Small*, **2012**, *8*, 2647–2651.

[178] Migahed, M., and Zidan, H. Influence of UV-irradiation on the structure and optical properties of polycarbonate films, *Curr. Appl Phys.*, **2006**, *6*, 91–96.

[179] Fritz, B., Hünig, R., Schmager, R., Hetterich, M., Lemmer, U., and Gomard, G. Assessing the influence of structural disorder on the plant epidermal cells optical properties: a numerical analysis, *Bioinspir. Biomim.*, **2017**, *12*, 036011–036020.

[180] Huld, T., Šúri, M., and Dunlop, E. D. Comparison of potential solar electricity output from fixed-inclined and two-axis tracking photovoltaic modules in Europe, *Prog. Photovoltaics Res. Appl.*, **2008**, *16*, 47–59.

[181] Šúri, M., Huld, T. A., Dunlop, E. D., and Ossenbrink, H. A. Potential of solar electricity generation in the European Union member states and candidate countries, *Sol. Energy*, **2007**, *81*, 1295–1305.

[182] Jung, Y. C., and Bhushan, B. Dynamic Effects of Bouncing Water Droplets on Superhydrophobic Surfaces, *Langmuir*, **2008**, *24*, 6262–6269.

[183] Tsai, P., Pacheco, S., Pirat, C., Lefferts, L., and Lohse, D. Drop Impact upon Micro- and Nanostructured Superhydrophobic Surfaces, *Langmuir*, **2009**, *25*, 12293–12298.

[184] Bartolo, D., Bouamrirene, F., Verneuil, E., Buguin, A., Silberzan, P., and Moulinet, S. Bouncing or sticky droplets: Impalement transitions on superhydrophobic micropatterned surfaces, *EPL (Europhysics Letters)*, **2006**, *74*, 299–305.

[185] Reyssat, M., Pépin, A., Marty, F., Chen, Y., and Quéré, D. Bouncing transitions on microtextured materials, *EPL (Europhysics Letters)*, **2006**, *74*, 306–312.

[186] Pye, K. *Aeolian dust and dust deposits*; Elsevier, **2015**.

[187] Itani, G. N., and Smith, C. A. Dust rains deliver diverse assemblages of microorganisms to the Eastern Mediterranean, *Sci. Rep.*, **2016**, *6*, 22657–22667.

[188] Byers, H. R. *General meteorology*; McGraw-Hill, **1959**.

[189] Luo, W., Khoo, Y. S., Hacke, P., Naumann, V., Lausch, D., Harvey, S. P., Singh, J. P., Chai, J., Wang, Y., Aberle, A. G., and Ramakrishna, S. Potential-induced degradation in photovoltaic modules: a critical review, *Energy Environ. Sci.*, **2017**, *10*, 43–68.

Glossary

Ca	capillary number	28
E	surface energy	6
F_E	embossing force	17
γ	surface tension	5
γ_W	water surface tension	61
I_{SC}	short-circuit current	104
J_{SC}	short-circuit current density	32
p_{air}	retained air pressure	10
p_{amb}	ambient air pressure	10
p_{base}	base pressure	68
p_{cap}	capillary pressure	10
p_{cr}	critical pressure	10
ϕ_S	liquid solid contact fraction	8
p_{water}	hydrostatic pressure	10
T_D	demolding temperature	37
T_E	embossing temperature	17
T_g	glass transition temperature	13
θ_{adv}	advancing contact angle	9
θ_{OS}	oil sliding angle	49
θ_O	oil contact angle	49
θ_{rec}	receding contact angle	9
θ_{WS}	water sliding angle	9
θ_W	water contact angle	6
T_m	melting temperature	14

Acronyms

ADSA	axisymmetric drop shape analysis	23
AOI	angle of incidence	31
AR	aspect ratio	19
Ca	capillary number	28
CA	contact angle	6
CAH	contact angle hysteresis	9
CLSM	confocal laser scanning microscope	29
COC	cyclo-olefin-copolymer	14
DI	deionized	28
DS wrinkles	double scale teflon surface wrinkles	33
EQE	external quantum efficiency	31
FAS	1H,1H,2H,2H-perfluorodecyltriethoxysilane	56
FEP	fluorinated ethylene propylene	14
HD	high density nanofur	38
IR	infrared	14
I-V	current-voltage	104
LD	low density nanofur	38
MCA	microcavity array	33
mc-Si	multicrystalline silicon	32
MD	medium density nanofur	38
OLED	organic light emitting diode	92
PC	polycarbonate	13
PDMS	polydimethylsiloxane	15
PE	polyethylene	46
p-nanofur	plasma treated nanofur	49
pp-nanofur	pressure-supported perforated nanofur	63
PTFE	polytetrafluorethylene	33

PV	photovoltaic	2
SEM	scanning electron microscope	21
UV	ultraviolet	15

Acknowledgement

This work would not have been possible without the assistance and support from numerous people. All the effort, workforce, experience and support extended towards accomplishing this thesis is highly appreciated. I therefore hope that all the aid and support is completely and justly recognized in the following without leaving anyone unmentioned.
First of all, I would like to thank my supervisor Dr. Hendrik Hölscher for providing me with the opportunity to pursue my Ph.D. thesis. Thank you for guiding me throughout the work and always providing advice, while also giving me the possibility to follow my own ideas and grow in the process.
Next, I would like to thank Prof. Bettina Frohnapfel and Prof. Chiara Neto for, without hesitation, taking upon them the tasks and responsibilities that come along with being co-referees to my thesis.
Tremendous thanks goes to Maryna Kavalenka, for her excellent supervision, endless support and assistance during the work. Thank you for listening to all the problems and issues and providing a bountiful of advice on how to overcome all the obstacles I encountered during my Ph.D.
A special thanks goes to the Landesgraduiertenförderung and the German Academic Exchange Service (DAAD) for funding the research project as well as a research stay in Sydney. From the University of Sydney I want to thank the Neto research group for heartily welcoming me, despite my sore lack of proper chemistry knowledge. Special thanks goes to Sam Peppou-Chapman and Liam Scarratt, who, besides discussing scientific ideas and always lending a helping hand in the lab, also assisted with all the small things necessary for making my stay in Sydney an unforgettable time.

Thanks is also owed to the students I had the pleasure to supervise and work with, namely Yann Germain, Meike Kronast, Luce-Marie Petit and Pooja Thakkar. A special thanks also goes to Benjamin Fritz and Guillaume Gomard from the Light Technology Institute (LTI) for the productive and easy cooperation. Moreover, I would like to thank the whole staff at IMT for on-going support, advice, tips and tricks, without which, many things would have been impossible. To just name a few: Alexandra Moritz and her workshop team; Dr. Uwe Köhler and the clean room team; Dr. Matthias Worgull, Marc Schneider and Andreas Striegel for the never ending and fast support with everything related to hot embossing; Holger Moritz, Siegfrid Roth, Mr. and Mrs. Klein and Nina Giraud for tackling all the non-scientific related problems and issues that accompany a Ph.D. project.

Also thank you to my office mates Randy Fechner, Christian Lutz, Tobias Förtsch and Florian Brüderlin. Thank you for the fun talks (scientific or not), discussions about proper labeling of axis, rage induced outbursts about things that didn't work and all the other things, which made the office a good natured environment.

I am thankful for all the friends, that made my Ph.D. an unforgettable, exciting and enjoyable adventure. Frederik Kotz for our weekly rounds of jogging around the campus, which not only kept us moving but also provided much needed time to vent frustration and go on rants about diverse topics. Andreas Striegel for ensuring that not everything is taken too seriously and that there is always some humor left in sticky situations. Senta Schauer for providing comic relief and being both an ally and the occasional spy. Richard Thelen for being the good natured, optimistic person he is, who always sees and points out the silver lining. Moreover, I want to thank all the other Ph.D. students, Post-Docs and alumni for always providing support, insightful discussions and the occasional after work drink. A heartfelt thank you goes to Markus and Andrea Gudernatsch and of course their children, who made these last years more enjoyable through shared dinners, weekend trips and the occasional drink.

A special thanks goes to my wife Jutta Vüllers. Dear Jutta, I cannot begin to describe how thankful I am for all the things you have done for me. I am especially thankful for your unwavering support, motivation and encouragement during these last few months and through every decision I made. You endured every mood or indecision on my part without ever faltering and still found the strength and right words to cheer me back up.

Finally, I want to thank my family. First and foremost I thank my parents Hans and Margarete Vüllers for their unconditional love and support. You respected all my decisions, had my back in every situation and helped in innumerable ways. I am grateful for everything you made possible for me. I also want to thank my brothers Hannes Vüllers and Franz Vüllers and their families. A fair share of laughs, giggles, smirks, snarles and "battle scars" are due to being your brother. Still I always enjoyed spending time with you guys and am ever grateful to see my nieces and nephews grow up.

Schriften des Instituts für Mikrostrukturtechnik
am Karlsruher Institut für Technologie (KIT)

ISSN 1869-5183

Herausgeber: Institut für Mikrostrukturtechnik

Band 1 Georg Obermaier
Research-to-Business Beziehungen: Technologietransfer durch Kommunikation von Werten (Barrieren, Erfolgsfaktoren und Strategien). 2009
ISBN 978-3-86644-448-5

Band 2 Thomas Grund
Entwicklung von Kunststoff-Mikroventilen im Batch-Verfahren. 2010
ISBN 978-3-86644-496-6

Band 3 Sven Schüle
Modular adaptive mikrooptische Systeme in Kombination mit Mikroaktoren. 2010
ISBN 978-3-86644-529-1

Band 4 Markus Simon
Röntgenlinsen mit großer Apertur. 2010
ISBN 978-3-86644-530-7

Band 5 K. Phillip Schierjott
Miniaturisierte Kapillarelektrophorese zur kontinuierlichen Überwachung von Kationen und Anionen in Prozessströmen. 2010
ISBN 978-3-86644-523-9

Band 6 Stephanie Kißling
Chemische und elektrochemische Methoden zur Oberflächenbearbeitung von galvanogeformten Nickel-Mikrostrukturen. 2010
ISBN 978-3-86644-548-2

Band 7 Friederike J. Gruhl
Oberflächenmodifikation von Surface Acoustic Wave (SAW) Biosensoren für biomedizinische Anwendungen. 2010
ISBN 978-3-86644-543-7

Band 8 Laura Zimmermann
Dreidimensional nanostrukturierte und superhydrophobe mikrofluidische Systeme zur Tröpfchengenerierung und -handhabung. 2011
ISBN 978-3-86644-634-2

Band 9 Martina Reinhardt
Funktionalisierte, polymere Mikrostrukturen für die dreidimensionale Zellkultur. 2011
ISBN 978-3-86644-616-8

Band 10 Mauno Schelb
Integrierte Sensoren mit photonischen Kristallen auf Polymerbasis. 2012
ISBN 978-3-86644-813-1

Band 11 Daniel Auernhammer
Integrierte Lagesensorik für ein adaptives mikrooptisches Ablenksystem. 2012
ISBN 978-3-86644-829-2

Band 12 Nils Z. Danckwardt
Pumpfreier Magnetpartikeltransport in einem Mikroreaktionssystem: Konzeption, Simulation und Machbarkeitsnachweis. 2012
ISBN 978-3-86644-846-9

Band 13 Alexander Kolew
Heißprägen von Verbundfolien für mikrofluidische Anwendungen. 2012
ISBN 978-3-86644-888-9

Band 14 Marko Brammer
Modulare Optoelektronische Mikrofluidische Backplane. 2012
ISBN 978-3-86644-920-6

Band 15 Christiane Neumann
Entwicklung einer Plattform zur individuellen Ansteuerung von Mikroventilen und Aktoren auf der Grundlage eines Phasenübergangs zum Einsatz in der Mikrofluidik. 2013
ISBN 978-3-86644-975-6

Band 16 Julian Hartbaum
Magnetisches Nanoaktorsystem. 2013
ISBN 978-3-86644-981-7

Band 17 Johannes Kenntner
Herstellung von Gitterstrukturen mit Aspektverhältnis 100 für die Phasenkontrastbildgebung in einem Talbot-Interferometer. 2013
ISBN 978-3-7315-0016-2

Band 18 Kristina Kreppenhofer
Modular Biomicrofluidics - Mikrofluidikchips im Baukastensystem für Anwendungen aus der Zellbiologie. 2013
ISBN 978-3-7315-0036-0

Band 19 Ansgar Waldbaur
Entwicklung eines maskenlosen Fotolithographiesystems zum Einsatz im Rapid Prototyping in der Mikrofluidik und zur gezielten Oberflächenfunktionalisierung. 2013
ISBN 978-3-7315-0119-0

Band 20 Christof Megnin
Formgedächtnis-Mikroventile für eine fluidische Plattform. 2013
ISBN 978-3-7315-0121-3

Band 21 Srinivasa Reddy Yeduru
Development of Microactuators Based on the Magnetic Shape Memory Effect. 2013
ISBN 978-3-7315-0125-1

Band 22 Michael Röhrig
Fabrication and Analysis of Bio-Inspired Smart Surfaces. 2014
ISBN 978-3-7315-0163-3

Band 23 Taleieh Rajabi
Entwicklung eines mikrofluidischen Zweikammer-Chipsystems mit integrierter Sensorik für die Anwendung in der Tumorforschung. 2014
ISBN 978-3-7315-0220-3

Band 24 Frieder Märkle
Laserbasierte Verfahren zur Herstellung hochdichter Peptidarrays. 2014
ISBN 978-3-7315-0222-7

Band 25 Tobias Meier
Magnetoresistive and Thermoresistive Scanning Probe Microscopy with Applications in Micro- and Nanotechnology. 2014
ISBN 978-3-7315-0253-1

Band 26 Felix Marschall
Entwicklung eines Röntgenmikroskops für Photonenenergien von 15 keV bis 30 keV. 2014
ISBN 978-3-7315-0263-0

Band 27 Leonardo Pires Carneiro
Development of an Electrochemical Biosensor Platform and a Suitable Low-Impedance Surface Modification Strategy. 2014
ISBN 978-3-7315-0272-2

Band 28 Sebastian Mathias Schillo
Prozessentwicklung für die Automatisierung der Herstellung und Anwendung von hochdichten Peptidmicroarrays. 2014
ISBN 978-3-7315-0274-6

Band 29 Nicole E. Steidle
Micro- and Nanostructured Microfluidic Devices for Localized Protein Immobilization and Other Biomedical Applications. 2014
ISBN 978-3-7315-0297-5

Band 30 Jochen Heneka
Prozessentwicklung eines industrietauglichen Verfahrens zur Fertigung von vereinzelten LIGA-Mikrobauteilen. 2015
ISBN 978-3-7315-0326-2

Band 31 Seoung-Eun Kim
Konzeption und prototypische Fertigung einer nicht-invasiven mikrofluidischen Plattform für die Elektrophysiologie (NIMEP) zur Zellenanlyse. 2015
ISBN 978-3-7315-0378-1

Band 32 Elisabeth Wilhelm
Entwicklung eines mikrofluidischen Brailledisplays. 2015
ISBN 978-3-7315-0385-9

Band 33 Viktor Pinneker
Entwicklung miniaturisierter Aktorsysteme basierend auf magnetischen Formgedächtnislegierungen. 2018
ISBN 978-3-7315-0500-6

Band 34 Ali Caglar Özen
Novel MRI Technologies for Structural and Functional Imaging of Tissues with Ultra-short T_2 Values. 2017
ISBN 978-3-7315-0657-7

Band 35 Anne Bäcker
Veränderliche 3D Zellgerüstträger auf Cryogelbasis zur Kultivierung von Prostatakarzinomzellen. 2017
ISBN 978-3-7315-0676-8

Band 36 Frieder Johannes Koch
X-ray optics made by X-ray lithography: Process optimization and quality control. 2017
ISBN 978-3-7315-0679-9

Band 37 Tobias Jörg Schröter
Vergrößerung des Sehfeldes der Röntgen-Phasenkontrast-Bildgebung für die klinische Anwendung. 2017
ISBN 978-3-7315-0731-4

Band 38 Felix Vüllers
Bioinspired Superhydrophobic Nano- and Microstructured Surfaces for Drag Reduction and Optoelectronics. 2018
ISBN 978-3-7315-0816-8